93418

12-50
QA

UNIVE

Algebraic Numbers and Diophantine Approximation

PURE AND APPLIED MATHEMATICS

A Series of Monographs and Textbooks

EXECUTIVE EDITOR

Earl J. Taft

RUTGERS UNIVERSITY
NEW BRUNSWICK, NEW JERSEY

CHAIRMAN OF THE EDITORIAL BOARD

S. Kobayashi

UNIVERSITY OF CALIFORNIA AT BERKELEY

1. K. YANO. Integral Formulas in Riemannian Geometry (1970)
2. S. KOBAYASHI. Hyperbolic Manifolds and Holomorphic Mappings (1970)
3. V. S. VLADIMIROV. Equations of Mathematical Physics (A. Jeffrey, editor; A. Littlewood, translator) (1970)
4. B. N. PSHENICHNYI. Necessary Conditions for an Extremum (L. Neustadt, translation editor; K. Makowski, translator) (1971)
5. L. NARICI, E. BECKENSTEIN, and G. BACHMAN, Functional Analysis and Valuation Theory (1971)
6. D. S. PASSMAN. Infinite Group Rings (1971)
7. L. DORNHOFF. Group Representation Theory (in two parts). Part A: Ordinary Representation Theory. Part B: Modular Representation Theory (1971, 1972)
8. W. BOOTHBY and G. L. WEISS (eds.). Symmetric Spaces: Short Courses Presented at Washington University (1972)
9. Y. MATSUSHIMA. Differentiable Manifolds (E. T. Kobayashi, translator) (1972)
10. L. E. WARD, JR. Topology: An Outline for a First Course (1972)
11. A. BABAKHANIAN. Cohomological Methods in Group Theory (1972)
12. R. GILMER. Multiplicative Ideal Theory (1972)
13. J. YEH. Stochastic Processes and the Wiener Integral (1973)
14. J. BARROS-NETO. Introduction to the Theory of Distributions (1973)
15. R. LARSEN. Functional Analysis: An Introduction (1973)
16. K. YANO and S. ISHIHARA. Tangent and Cotangent Bundles: Differential Geometry (1973)
17. C. PROCESI. Rings with Polynomial Identities (1973)
18. R. HERMANN. Geometry, Physics, and Systems (1973)
19. N. R. WALLACH. Harmonic Analysis on Homogeneous Spaces (1973)
20. J. DIEUDONNÉ. Introduction to the Theory of Formal Groups (1973)
21. I. VAISMAN. Cohomology and Differential Forms (1973)
22. B.-Y. CHEN. Geometry of Submanifolds (1973)
23. M. MARCUS. Finite Dimensional Multilinear Algebra (in two parts) (1973)
24. R. LARSEN. Banach Algebras: An Introduction (1973)
25. R. O. KUJALA and A. L. VITTER (eds.). Value Distribution Theory: Part A; Part B, Deficit and Bezout Estimates by Wilhelm Stoll (1973)
26. K. B. STOLARSKY. Algebraic Numbers and Diophantine Approximation (1974)
27. A. R. MAGID. The Separable Galois Theory of Commutative Rings (1974)

Algebraic Numbers and Diophantine Approximation

KENNETH B. STOLARSKY
Department of Mathematics
University of Illinois
Urbana, Illinois

MARCEL DEKKER, INC. New York 1974

959168

Copyright © 1974 by MARCEL DEKKER, INC.

ALL RIGHTS RESERVED

Neither this book nor any part may be reproduced or transmitted in any form or by any means, electronic or mechanical, including photocopying, microfilming, and recording, or by any information storage and retrieval system, without permission in writing from the publisher.

MARCEL DEKKER, INC.
270 Madison Avenue, New York, New York 10016

LIBRARY OF CONGRESS CATALOG CARD NUMBER: 73-85382

ISBN: 0-8247-6102-2

Current printing (last digit):
10 9 8 7 6 5 4 3 2 1

PRINTED IN THE UNITED STATES OF AMERICA

הכל צפוי והרשות נתונה

ובטוב העולם נדון

והכל לפי רב המעשה

רבי עקיבא

CONTENTS

PREFACE

This volume is intended as an introduction to Diophantine approximation, that presupposes no more than a basic course in abstract algebra (say at the level of van der Waerden's _Modern Algebra_) and one in complex variables (at the level of Ahlfors' _Complex Variables_). Instead of trying to survey all the topics usually classified as Diophantine approximation, it concentrates on two great results, that of K. F. Roth on rational approximations to algebraic numbers and that of A. Baker on linear forms in the logarithms of algebraic numbers (in fact Roth and Baker have each received a Fields medal for their work). Both of these results are motivated by the perennial question: "How does one solve a Diophantine equation?"

Roth's method is an ingenious development of earlier ideas of A. Thue and of C. L. Siegel, and further important progress using these ideas has recently been made by E. Wirsing and by W. Schmidt. On the other hand, Baker's famous results were obtained by a surprising development of a method of A. Gelfond; further progress using this method has recently been made by Baker himself, by J. Coates, by N. I. Feldman, and by H. Stark.

I cannot emphasize too strongly that no part of this volume is intended to serve as a "who's who" of Diophantine approximation. To keep it short much that is excellent in this area has been omitted, while some material that points in other directions has been included to provide variety. Perhaps the most glaring omissions are Siegel's theory of E-functions and (aside from a faint glimmer in Exercise 7.5.1) the p-adic theory. For these the reader is referred to Siegel [110] and Mahler [67]. The latter treats Ridout's theorem and similar topics. Another omission is Skolem's method; an exposition can be found in Borevich and Shafarevich [13]. Also, no attempt is made to provide a complete bibliography. The interested reader can find a wide list of references in the excellent survey papers of Feldman and Shidlovskii [35], Lang [55], and Schmidt [101], and in other books on Diophantine approximation. Some very important new results (e.g., the n-dimensional Roth theorem) appear in the 1970 University of Colorado notes Lectures on Diophantine Approximation by Schmidt [100].

My task has been largely one of arrangement and exposition, and it will be clear to any expert how very much I owe to previous books and papers. However, I do think there is some novelty in the presentation of the extrapolation technique of Section 8.4, and in the formulation of Proposition 11.2.1. Also, to put Baker's results in an appropriate setting (they often seem to be weaker than Roth's results, but in distinction to Roth's results they are constructive), I develop the rudiments of algebraic number theory from a constructive point of view. This is done somewhat more systematically than in Borevich and Shafarevich [13], for example. The reader who would like to know more about which parts of mathematics can be made constructive (an open problem in general) should consult Bishop [11] and Tarski [121].

Wide areas of our subject are touched on only in the exercises, but here I have often provided rather detailed hints. The reader would do well to at least skim the exercises in order to get a broader view of the subject. In Sections 3 and 4 I have provided an overabundance of propositions and exercises on the heights and lengths of algebraic numbers; I feel that this subject ought to be systematized and hope that I have provided some impetus toward this goal.

The exercises preceded by an asterisk (*) are intended to be more difficult than those that are not. A few of these exercises are quite hard, and I am generally unable to work those with two asterisks.

I wish to thank Professor Wolfgang Schmidt of the University of Colorado and Professor Eduard Wirsing of the Mathematisches Institut, Marburg, for generously allowing me to make use of some of their unpublished manuscripts, and Mr. Clifton Corzatt of Urbana, Illinois, for proofreading parts of the first four chapters. Professor Schmidt also contributed several useful suggestions.

Finally, I thank my typist, Mrs. Divona Wrather, whose care in preparing this manuscript always matched and often exceeded what could reasonably be expected from even the finest technical typist.

RECURRENT NOTATIONS

α, α_i	algebraic numbers
$\underset{\sim}{A}$	field of all algebraic numbers
β, β_i; γ, γ_i	algebraic numbers
B	upper bound
c, c_i; C, C_i	positive constants independent of main parameters
$\underset{\sim}{C}$	field of complex numbers
$\mathcal{C}$	contour of integration
$d_x(P)$	degree of the polynomial P in the variable x
d(P)	degree of the one-variable polynomial P
$d(\alpha)$	degree of the algebraic number α
$d(\underset{\sim}{K})$	degree of the algebraic number field $\underset{\sim}{K}$
deg Δ	degree of the differential operator Δ
$d(\Lambda)$	determinant of the lattice Λ
det(M)	determinant of the matrix M
$\underset{\sim}{F}$	field
$\underset{\sim}{F}(x)$	field of all rational functions in x with coefficients in $\underset{\sim}{F}$
$\underset{\sim}{F}[x]$	ring of polynomials in x with coefficients from $\underset{\sim}{F}$
$\underset{\sim}{F}[x_1,\ldots,x_n]$	ring of polynomials in $x_1,\ldots,x_n$ with coefficients from $\underset{\sim}{F}$
$\underset{\sim}{F}(\alpha)$	the finite field extension of $\underset{\sim}{F}$ obtained by adjoining the algebraic number α

$\underset{\sim}{F}(\alpha_1,\ldots,\alpha_n)$	the finite field extension of $\underset{\sim}{F}$ obtained by adjoining the algebraic numbers $\alpha_1,\ldots,\alpha_n$
$f(x,y) = m$	a two-variable Diophantine equation
$H(P)$	height of the polynomial P
$H(\alpha)$	height of the algebraic number α
$I(P)$	index of the polynomial P
$Im(z)$	imaginary part of the complex number z
$\underset{\sim}{K}$	an algebraic number field
$\underset{\sim}{K}/\underset{\sim}{F}$	the field $\underset{\sim}{K}$ considered as an extension of the field $\underset{\sim}{F}$
$L(P)$	length of the polynomial P
Λ	an n-dimensional lattice
$N(\alpha)$	norm of the algebraic number α
ω_i	algebraic numbers, especially elements of an integral basis
$p, p_i; q, q_i$	rational integers
P, Q	polynomials
$P(x)$	polynomial in x
$P(x_1,\ldots,x_n)$	polynomial in n variables $x_1,\ldots,x_n$
$\underset{\sim}{Q}$	field of rationals
$Re(z)$	real part of the complex number z
$\underset{\sim}{R}$	field of real numbers
$\underset{\sim}{R}^n$	real n-dimensional space
S	set
σ, τ	elements of Galois groups

$\mathrm{Tr}(\alpha)$	trace of the algebraic number α
t	real parameter
$\underline{v}, \underline{w}, \underline{x}, \underline{y}$	vectors
ξ	a real or complex parameter
$\underset{\sim}{Z}$	set of all integers
$\underset{\sim}{Z}^{+}$	set of all positive integers

Algebraic Numbers and Diophantine Approximation

Chapter 1

DIOPHANTINE EQUATIONS

To most people there are no abstract concepts simpler than the addition and multiplication of positive integers. Yet even here "a fool can ask a question that a thousand wise men cannot answer." For example, which positive integers x,y satisfy

$$y^3 = x^2 + 999? \tag{1.1}$$

Equation (1.1) is an example of what is called a Diophantine equation, although the Babylonians had considered such equations long before Diophantus (see Neugebauer [74] and Neugebauer and Sachs [75]).

Definition 1.1. A polynomial Diophantine equation is an equation of the form

$$P(x_1,\ldots,x_m) = 0, \tag{1.2}$$

where $m \in \underset{\sim}{Z}^+$ and $P \in \underset{\sim}{Z}[x_1,\ldots,x_m]$; that is, m is a positive integer and P is a polynomial in m variables with integer coefficients. The x_i, $i = 1,\ldots,m$, are understood to be integer-valued variables.

If $P(a_1,\ldots,a_m) = 0$, where $a_i \in \underset{\sim}{Z}$ for $i = 1,\ldots,m$, we say that $(a_1,\ldots,a_m)$ is a solution of (1.2). For example, (1,10) is a solution ($x_1 = x$, $x_2 = y$) to (1.1), whereas $x_1^2 + x_2^2 + x_3^2 = 7$

has no solutions.

The following basic question is known as Hilbert's tenth problem: "Is there an algorithm for determining whether or not any given polynomial Diophantine equation has a solution?" Here we assume that the reader understands intuitively what is meant by an "algorithm." A precise definition can be found in most books on mathematical logic (see, for example, Davis [25], Introduction and Chapter 1).

The first approach to Hilbert's tenth problem was found by J. Robinson [89]. Her work led to the result of Davis, Putman, and J. Robinson [27] that if we extend the class of Diophantine equations to include the so-called exponential Diophantine equations,

$$P(x_1,\ldots,x_m;2^{x_1},\ldots,2^{x_m}) = 0,$$

then no algorithm can exist. In 1970 Matijasević [73] perfected J. Robinson's method and proved that the answer to Hilbert's original problem is negative. In fact (see Davis [26] for an excellent exposition) there is a polynomial $Q(y,z,w;x_1,\ldots,x_m)$ such that $y = z^w$ if and only if there exist positive integers $x_1,\ldots,x_m$ such that $Q(y,z,w;x_1,\ldots,x_m) = 0$, so with a little work it can be shown that exponential Diophantine equations are no harder than polynomial Diophantine equations.

Note, however, that if an "oracle" tells us that a particular polynomial Diophantine equation has solutions, we can always find one simply by enumerating all m-tuples of natural numbers and testing them one at a time.

Henceforth by a Diophantine equation we shall mean a polynomial Diophantine equation. More specifically we shall say that it is of degree n in m variables and of height H if P is a polynomial of degree n in m variables and H is the maximum of the absolute values of the coefficients of P. For example, (1.1) is of degree 3 in two variables and of height 999.

Instead of Hilbert's problem we may ask more modestly for algorithms that solve special classes of Diophantine equations, though in these cases we might less modestly require that our algorithm produce all solutions. For Diophantine equations in two variables of degree n this has been done for n = 1,2,3 respectively by the ancient Greeks, C. F. Gauss, and A. Baker (better: algorithms exist for two-variable Diophantine equations of genus $g \leq 1$). Alternatively we could ask for an upper bound on the number of solutions or merely whether or not the number is infinite (see Section 6 of Gelfond [38] for a summary of what is known and Siegel [108] or [111], pp. 242-246, for the actual proofs). However, the value of finding only some solutions of a fixed Diophantine equation is usually rather small (see Dickson [28], p xx).

At first one might think that a knowledge of the integers alone should suffice for the study of Diophantine equations. This might be true in some logical sense, but the only way of making the subject humanly accessible seems to be to embed the integers $\mathbb{Z}$ in the rationals $\mathbb{Q}$, then $\mathbb{Q}$ in appropriate algebraic number fields $\mathbb{Q}(\alpha_i)$, and these in various metric completions and algebraic closures. (Contrast the treatment of $x^3 + y^3 = z^3$ in LeVeque [58], pp. 95-97, with the completely elementary approach in Sierpinski [113], pp. 384-388.) For example, say we wish to solve the homogeneous (i.e., each term having the same degree) Diophantine equation

$$P(x,y) = a_n x^n + a_{n-1} x^{n-1} y + \cdots + a_n y^n = m, \tag{1.3}$$

where $P(x,y)$ is irreducible, $a_i, m \in \underset{\sim}{Z}$ for $i = 1,\ldots,n$, and $a_n, m \neq 0$. In the field of complex numbers $\underset{\sim}{C}$, Gauss's fundamental theorem of algebra allows us to write this as

$$\left(\frac{x}{y} - \alpha_1\right) \cdots \left(\frac{x}{y} - \alpha_n\right) = \frac{m}{a_n y^n}, \tag{1.4}$$

where $\alpha_i \in \underset{\sim}{C}$ for $i = 1,\ldots,n$. (More precisely, the α_i are distinct algebraic numbers of degree n over $\underset{\sim}{Q}$.) Now $m/a_n y^n \to 0$ as $|y| \to \infty$, and for fixed y there are at most n values of x such that (x,y) is a solution. Hence, if solutions are (partially) ordered by the size of $|y|$, large solutions must correspond to good approximations of some α_i by rational fractions x/y. However, we shall see that $x,y \in \underset{\sim}{Z}$ and $y \neq 0$ imply that

$$\left|\frac{x}{y} - \alpha\right| > \frac{\theta(y,n,\alpha)}{y^n}, \tag{1.5}$$

where $\theta(y,n,\alpha) \to \infty$ as $y \to \infty$, for any algebraic α (see Section 3.1 for the definition of "algebraic"). Also, for some $\varepsilon > 0$ and $|y|$ sufficiently large,

$$\left|\frac{x}{y} - \alpha_j\right| > \varepsilon \quad \text{if} \quad j \neq i = i(y),\ 1 \leq i,j \leq n.$$

Hence from (1.4),

$$\left|\frac{m}{a_n}\right| \frac{1}{y^n} > \frac{\epsilon^{n-1}}{y^n} \min_{i=1}^{n} \theta(y,n,\alpha_i),$$

from which it follows that there are only finitely many solutions to (1.3). Note, however, that the mere existence of the $\theta(y,n,\alpha_i)$ does not enable us to decide whether or not (1.3) has a solution.

Diophantine approximation consists of the study of inequalities of the same general type as (1.5).

Exercises

Let $\mathcal{P}$ be the set of all polynomials in any number of variables with integer coefficients. For $P = P(x_1,\ldots,x_m) \in \mathcal{P}$ define functions f,g on $\mathcal{P}$ by

$$f(p) = \begin{cases} 0 \text{ if } P(x_1,\ldots,x_n) = 0 \text{ has no solution with } x_i \in \mathbb{Z} \\ 1 \text{ otherwise} \end{cases}$$

and similarly for g, with $\mathbb{Z}^+$ in place of $\mathbb{Z}$.

1.1. Show that $f(ax + by + c) = \begin{cases} 0 \text{ if } (a,b) \nmid c \\ 1 \text{ otherwise.} \end{cases}$

1.2. Show that there is an algorithm for computing f if and only if there is one for g.

1.3. Show that any Diophantine equation is equivalent to (a) a system of quadratic Diophantine equations; (b) a single quartic Diophantine equation. (Hint: See Skolem [114], pp. 2 and 3.)

Let $P = P(x_1,\ldots,x_n;y_1,\ldots,y_n)$ be a polynomial in $(n + m)$ integer-valued variables. A statement of the form "$P = 0$" is called a *polynomial predicate* and defines a (possibly empty) subset of $\mathbf{Z}^{n+m}$. A statement of the form "There exist $x_1,\ldots,x_n \in \mathbf{Z}$ such that $P(x_1,\ldots,x_n;y_1,\ldots,y_m) = 0$" is called a *Diophantine predicate* and defines a subset of $\mathbf{Z}^m$.

1.4. Show that if A,B are polynomial predicates, then so are "A and B" and "A or B."

1.5. Show that if $A(z;y_1,\ldots,y_m)$ is a polynomial predicate, then so are $B = B(x;y_1,\ldots,y_m)$ and $C = C(y_1,\ldots,y_m)$, where

$B \equiv$ "For all $z \in \{0,1,2,\ldots,x\}$ the predicate $P(z;y_1,\ldots,y_m)$ is true."

$C \equiv$ "For all $z \in \mathbf{Z}$ the predicate $P(z;y_1,\ldots,y_m)$ is true."

1.6. Show that if A,B are Diophantine predicates, then so are "A and B" and "A or B."

We say that $S \subseteq \mathbf{Z}$ is *recursively enumerable* if $S = \{f(n) \mid n \in \mathbf{Z}\}$ for some number-theoretic function f and we have an algorithm for computing f at any integer n; we say that it is *recursive* if we have an algorithm for determining whether or not $m \in S$ for any $m \in \mathbf{Z}$. It is known that there are recursively enumerable sets that are not recursive. We call an $S \subseteq \mathbf{Z}$ *Diophantine* if $S = \{y \mid P(x_1,\ldots,x_n;y) = 0\}$. Matijasević's result on Hilbert's tenth problem can be formulated as follows: "Every recursively enumerable set is Diophantine."

For the original statement of Hilbert's tenth problem, see Hilbert [46,47].

Chapter 2

THE INTEGERS

2.1. Axiomatics

Basic to all mathematics is the axiom of induction:

(2.1.1) If S is a set of positive integers, $1 \in S$, and if $(n + 1) \in S$ whenever $n \in S$, then S is the set of all positive integers.

An equivalent formulation that is often more convenient is the following:

(2.1.2) A nonempty set S of nonnegative (or positive) integers has a minimal element.

The contrapositive form of (2.1.2), when used in a proof, is called "the method of infinite descent."

Although one can attempt to base much of mathematics almost entirely on an induction axiom (see Landau [53] or Cohen [20], pp. 20-25), it will be convenient here to think of it as based on the usual (naive) set theory. In particular, it is not immediately clear whether the final step in the proof of Wirsing's theorem (see Section 7.5) requires the axiom of choice and/or the axiom of infinity (in fact, the answer is negative in both cases). Also, Cauchy's residue theorem is used in Chapter 9, and this (in its full generality, which we do not require) is sometimes deduced from the axiom of choice via the Hahn-Banach theorem (Rudin [95], pp. 259 and 260).

A simple consequence of induction, the pigeonhole principle, can be stated quite vividly:

(2.1.3) If $(n + 1)$ objects are placed in n boxes, some box will contain at least two objects.

Another simple but important consequence of induction is discreteness: there is no integer between 0 and 1. This is most usefully stated in the case where the integers have already been embedded in $\mathbb{R}$ or $\mathbb{C}$:

(2.1.4) If $x \in \mathbb{Z}$ and $|x| < 1$, then $x = 0$.

Exercises

2.1.1. Use (2.1.1) to show that there are infinitely many solutions to the Diophantine equation $x^2 - 2y^2 = 1$. (Hint: If (x_n, y_n) is a solution, so is (x_{n+1}, y_{n+1}), where $x_{n+1} = 3x_n + 4y_n$ and $y_{n+1} = 2x_n + 3y_n$.)

2.1.2. Use (2.1.2) to derive the division algorithm, then the Euclidean algorithm, and hence an algorithm for finding all solutions of the Diophantine equation $ax + by + c = 0$, where $a,b,c \in \mathbb{Z}$ are fixed.

*2.1.3. Use infinite descent to show there do not exist $x,y,z \in \mathbb{Z}^+$ such that $x^4 + y^4 = z^4$. (Hint: See Hardy and Wright [44], pp. 191 and 192.)

2.1.4. Use (2.1.3) to show that for $\xi \in \underset{\sim}{R}$ and $m \in \underset{\sim}{Z}^+$ the simultaneous inequalities $|x - \xi y| < 1/m$, $0 < y < m$, have an integer solution (x,y).

2.1.5. Use (2.1.4) to show that e (the base of the natural logarithms) is irrational. (Hint: From $p/q = 1/e = \Sigma_{n=0}^{\infty}(-1)^n/n!$ it follows that $|p/q - m/n!| < 1/(n + 1)!$ for some $m \in \underset{\sim}{Z}^+$.)

2.1.6. Show that (2.1.3) is the case r = 1 of Ramsey's theorem: Let $\mathcal{P}_r$ be the set of all r-element subsets of S and say $\mathcal{P}_r = Q_1 \cup \cdots \cup Q_t$, where $Q_i \cap Q_j = \phi$, $i \neq j$. Write $T \overset{r}{\rightarrow} Q_i$ whenever every r-element subset of T ($\subseteq$ S) is an element of Q_i. Then for every $q \in \underset{\sim}{Z}^+$ there is an $N = N(q,r,t) \in \underset{\sim}{Z}^+$ such that $|S| \geq N$ implies that for some $T \subseteq S$ with $|T| = q$, we have $T \overset{r}{\rightarrow} Q_k$ for some $k \in [1,t]$. (Note: For a proof of Ramsey's theorem see Percus [82], pp. 42-47 , or Ramsey [87].)

2.1.7. Is the construction of $\underset{\sim}{R}$ dependent on the axiom of choice and/or the axiom of infinity? (Answer: No and yes, respectively.)

For more on the axiom of infinity and the problem of developing $\underset{\sim}{Z}$, $\underset{\sim}{Q}$, and $\underset{\sim}{R}$ from set theory, see Suppes [120], pp. 127-158, especially pp. 138 and 139. Also see p. 186, where Suppes carefully avoids using the axiom of choice.

2.2. Applications of the Pigeonhole Principle

A simple corollary of (2.1.3) is the following:

Corollary 2.2.1. If the cardinality of the domain of a function f exceeds that of its range, then we can find distinct elements x_1, x_2 in its domain such that $f(x_1) = f(x_2)$.

From Corollary 2.2.1 and (2.1.3) we obtain results on linear forms with integer-valued variables that are of fundamental importance in Diophantine approximation.

Lemma 2.2.1. Let M,N denote integers, $N > M > 0$, and let a_{ij} $(1 \le i \le M,\ i \le j \le N)$ denote integers not all zero with absolute values at most H. Then the system of linear Diophantine equations

$$(2.2.1) \qquad y_i = y_i(\underline{x}) = \sum_{j=1}^{N} a_{ij}x_j = 0 \qquad (1 \le i \le M)$$

has a nonzero solution $\underline{x} = (x_1,\ldots,x_N)$ with

$$(2.2.2) \qquad |x_j| \le (NH)^{M/(N-M)} \qquad (1 \le j \le N).$$

Note: $N > M$ guarantees a rational (hence an integral) solution, but we want a bound on the size of the solution.

Proof. Let $B = [(NH)^{M/(N-M)}]$. Then there are $(B + 1)^N$ vectors $\underline{x} = (x_1,\ldots,x_N)$ that have nonnegative components and satisfy (2.2.2). Consider them as the domain of the vector-valued function $\underline{y} = (y_1,\ldots,y_M)$. To estimate (the cardinality of) the range of $\underline{y}$, let L_i^+ (L_i^-) be the sum of the positive (negative) a_{ij}, $1 \le j \le N$. Then $L_i^- B \le y_i \le L_i^+ B$ and $L_i^+ - L_i^- \le NH$, so there are at most $(NBH + 1)^M$ possibilities for $\underline{y}$. But

$$(B + 1)^N > (NBH + 1)^M$$

since $(B + 1)^{N-M} > (NH)^M$, so for some pair of distinct vectors $\underline{x}_1, \underline{x}_2$ the corresponding $\underline{y}$ vectors are the same by Corollary 2.2.1.

Hence $\underline{x} = \underline{x}_1 - \underline{x}_2$ is the desired solution. □

Lemma 2.2.2. Let M,N denote integers, $N > M > 0$, and let a_{ij} $(1 \le i \le M,\ 1 \le j \le N)$ denote real numbers, not all zero, satisfying

$$\sum_{j=1}^{N} |a_{ij}| \le L_i \qquad (1 \le i \le M).$$

Then for a positive integer B we can find a nonzero integral $\underline{x} = (x_1,\ldots,x_N)$ such that

$$|y_i(\underline{x})| \le 2L_iB/\{(B+1)^{N/M} - 2\} \qquad (1 \le i \le M) \tag{2.2.3}$$

and

$$|x_j| \le B \qquad (1 \le j \le N). \tag{2.2.4}$$

Proof. For each $\underline{x}$ with $0 \le x_j \le B$, $j = 1,\ldots,N$, we have $|y_i(\underline{x})| \le L_iB$. As before, consider $\underline{y}$ as a function on the set of all $(B+1)^N$ such $\underline{x}$ vectors. It maps them into an M-dimensional rectangle of sides $2L_iB$, $i = 1,\ldots,M$. Subdivide this rectangle into n^M congruent subrectangles of sides $2L_iB/n$, $i = 1,\ldots,M$ (Fig. 1). By (2.1.3) two distinct vectors $\underline{x}_1,\underline{x}_2$ are mapped into

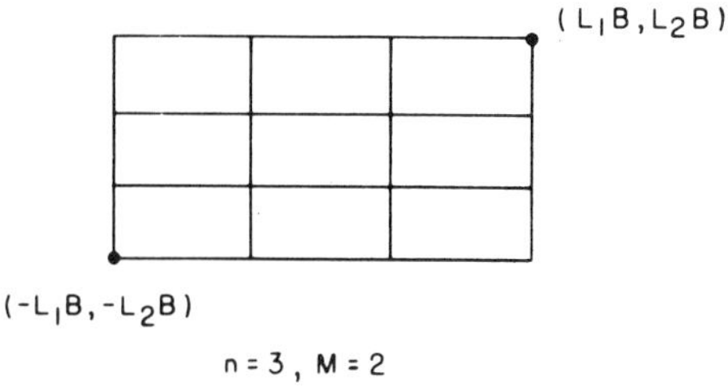

Figure 1

the same subrectangle if $(B + 1)^N > n^M$, hence if

$$n = [(B + 1)^{N/M} - 1].$$

Thus

$$|y_i(\underline{x}_1 - \underline{x}_2)| = |y_i(\underline{x}_1) - y_i(\underline{x}_2)| \leq \frac{2L_i B}{\{(B + 1)^{N/M} - 2\}},$$

and the components of $\underline{x}_1 - \underline{x}_2$ clearly satisfy (2.2.4). □

Exercises

Let α be algebraic of degree n over $\underset{\sim}{Q}$ (see Definitions 3.1.1 and 3.1.2) and let $\alpha = \alpha^{(1)}, \ldots, \alpha^{(n)}$ be its conjugates. Set

$$\underset{\sim}{K} = \underset{\sim}{Q}(\alpha) = \{r_0 + r_1\alpha + \cdots + r_{n-1}\alpha^{n-1} | r_i \in \underset{\sim}{Q},\ i = 1, \ldots, n\}.$$

For $\beta \in \underset{\sim}{K}$, let $\beta^{(k)} = r_0 + r_1\alpha^{(k)} + \cdots + r_{n-1}\alpha^{(k)n-1}$. Now say there are $\omega_1, \ldots, \omega_n$ in $\underset{\sim}{K}$ such that

(i) $\det[\omega_i^{(k)}] \neq 0$,

(ii) $\omega_i\omega_j = \sum_{k=0}^{n} c_{ijk}\omega_k \qquad c_{ijk} \in \underset{\sim}{Z}$.

Call $\beta \in \underset{\sim}{K}$ an ω-integer if $\beta = \sum_{i=0}^{n} b_i\omega_i$, where $b_i \in \underset{\sim}{Z}$.

<u>2.2.1</u>. Let M,N denote integers $N > M > 0$ and let $a_{ij} \in \underset{\sim}{K}$ $(1 \leq i \leq M,\ 1 \leq j \leq N)$ denote ω-integers not all zero with $|a_{ij}^{(k)}| \leq H$, $k = 1, \ldots, n$. Show that the system of linear equations

$$Y_i = Y_i(\underline{X}) = \sum_{j=1}^{N} a_{ij}X_j = 0 \qquad (1 \leq i \leq M)$$

has a nonzero solution $\underline{X} = (X_1, \ldots, X_N)$, where X_i is an ω-integer,

$$|X_j^{(k)}| \leq c_1(c_2NH)^{M/(N-M)} \qquad (1 \leq j \leq N,\ 1 \leq k \leq n),$$

and C_1, C_2 depend only on the ω_i, $i = 1,\dots,n$. (Hint: Let $C_i = C_i(\omega_1,\dots,\omega_n) > 0$. Write $a_{ij} = b_{ij1}\omega_1 + \cdots + b_{ijn}\omega_n$ (*) and $X_j = x_{j1}\omega_1 + \cdots + x_{jn}\omega_n$, where $b_{ijk} \in \mathbb{Z}$ and x_{jk} is an integer-valued variable. Then we can make $Y_i = 0$ by choosing the x_{jk} to satisfy a system $\mathcal{D}_i$ of n linear Diophantine equations in nN unknowns. From (*) we can derive a system of n equations in the unknowns b_{ijk}, $k = 1,\dots,n$, by taking conjugates. By assumption (i) we can solve them for the b_{ijk} in terms of the $a_{ij}^{(k)}$, and hence

$$|b_{ijk}| < C_3 H.$$

Hence the (integral) coefficients of the systems $\mathcal{D}_i$, $i = 1,\dots,M$, are all bounded by C_4H; so Lemma 2.2.1 applies to the simultaneous system $\mathcal{D} = \bigcup_{i=1}^{M} \mathcal{D}_i$ and

$$|x_{jk}| \le (nNC_4H)^{nM/(nN-nM)} \qquad (1 \le j \le N,\ 1 \le k \le n).$$

Thus

$$|X_j^{(k)}| < nC_5(nNC_4H)^{M/(N-M)}.)$$

Such $\omega_1,\dots,\omega_n$ always exist (e.g., take any integral basis of $\mathbb{K}$). For more details, see Gelfond [39], pp. 52 and 53.

<u>2.2.2</u>. Say $|c_{ij}| \le C$ and $|\det c_{ij}| \ge \delta$, where $c_{ij} \in \mathbb{C}$, $i,j = 1,\dots,n$.

(a) Find an upper bound for $|b_j|$, $j = 1,\dots,n$, if $|a_i| \le A$ for $i = 1,\dots,n$ and $a_i = \Sigma_{j=1}^{n} c_{ij}b_j$.

(b) Find explicit formulas for the C_1, C_2 of Exercise 2.2.1 in terms of $\delta = |\det[\omega_i^{(k)}]|$, $C = \max|c_{ijk}|$, $\omega = \max|\omega_i^{(k)}|$, and n.

2.2.3. Change the hypotheses of Lemma 2.2.2 by allowing the a_{ij} to be complex and requiring that $N > 2M$. Show that the conclusion, with M and L_i replaced by $2M$ and $2L_i$, respectively, in (2.2.3), is still valid.

2.2.4. Let $b_i > 0$ for $i = 1,\dots,N$ and in the hypothesis of Lemma 2.2.2 let $N = M$. If

$$b_1 \cdots b_N > L_1 \cdots L_N,$$

there is a nonzero integral $\underline{x} = (x_1,\dots,x_N)$ such that

$$|y_i(\underline{x})| < b_i \qquad (1 \leq i \leq N).$$

(Hint: For each $\underline{x}$ with $-B \leq x_i \leq B$, $i = 1,\dots,N$, we have $|y_i(\underline{x})| \leq L_i B$. As before, consider $\underline{y}$ as a function on the set of all $(2B + 1)^N$ such $\underline{x}$ vectors. It maps them into an N-dimensional rectangle $\mathcal{R}$ of sides $2L_iB$, $i = 1,\dots,N$. Subdivide the i-th side of $\mathcal{R}$ into $n_i = (2L_iB/b_i) + 1$ equal parts and extend this to a subdivision of $\mathcal{R}$ into $n_1 \cdots n_N$ subrectangles. For B sufficiently large,

$$n_1 \cdots n_N < (2B + 1)^N,$$

so two distinct vectors $\underline{x}_1,\underline{x}_2$ are mapped into the same subrectangle. Thus

$$|y_i(\underline{x}_1-\underline{x}_2)| \leq |y_i(\underline{x}_1) - y_i(\underline{x}_2)| < \frac{2L_iB}{(2L_iB)/b_i} = b_i$$

for $1 \leq i \leq N$. See Mahler [67], pp. 48-50.)

Theorem 11.2.2 leads to stronger results of this nature.

2.2.5. If $s_i, t_i \geq 0$, show that

$$\prod_{i=1}^{M} (s_i + t_i)^{1/M} \geq \prod_{i=1}^{M} x_i^{1/M} + \prod_{i=1}^{M} y_i^{1/M}.$$

(Hint: Divide through by the left-hand side and apply the inequality of the arithmetic-geometric mean.)

2.2.6. Let M,N denote integers with $N \geq M > 0$ and let the a_{ij} $(1 \leq i \leq M,\ 1 \leq j \leq N)$ denote real numbers, not all zero, satisfying

$$\sum_{j=1}^{N} |a_{ij}| < L_i \qquad (1 \leq i \leq M).$$

For the positive integers B_i $(1 \leq i \leq N)$ we can find a nonzero integral $\underline{x} = (x_1,\dots,x_N)$ such that $|x_i| \leq B_i$ and the

$$y_i = y_i(\underline{x}) = \sum_{j=1}^{N} a_{ij}x_j \qquad (1 \leq i \leq M)$$

satisfy

$$|y_i(\underline{x})| \leq 2L_i B/(B_1 \cdots B_N)^{1/M},$$

where $B = \max_{i=1}^{N} B_i$. (Hint: Follow the proof of Lemma 2.2.2 and use Exercise 2.2.5 to show there is an integer n satisfying

$$\prod_{i=1}^{N} (B_i + 1)^{1/M} \geq n \geq \prod_{i=1}^{N} B_i^{1/M}.)$$

Exercise 2.2.6 gives us an improved formulation of Lemma 2.2.2. In the case M = 1 it asserts that, given real numbers $a_0,\dots,a_N$, there are integers $x_0,\dots,x_N$ with $|x_i| \leq H_i$ such that

$$(*) \qquad |x_0a_0 + \cdots + x_Na_N| \leq \frac{2LH}{H_0 \cdots H_N},$$

where $L = |a_0| + \cdots + |a_N|$ and $H = \max_{i=0}^{N} H_i$. The problem of determining when (*) is in some sense best possible has only been studied quite recently. At the cost of getting very much ahead of ourselves, we point out the following special case of a theorem of Baker's [4]:

Theorem. Let e be the base of the natural logarithms. Then there is a $c = c(N) > 0$ such that

$$|x_0 + x_1 e + \cdots + x_N e^N| > \frac{cH^{1-\varepsilon(H)}}{H_0 \cdots H_N}$$

for any nonzero integers $x_0, \ldots, x_N$ with $|x_i| \leq H_i$, where

$$\varepsilon(t) = [c \log \log(1 + t)]^{-1/2}$$

and $H = \max_{i=1}^{N} H_i$.

*2.2.7. (A problem of Leo Moser's). Can Lemma 2.2.1 be strengthened so that (2.2.2) becomes

$$|x_j| \leq (cH)^{M/(N-M)}$$

where c is an absolute constant?

Chapter 3

ALGEBRAIC NUMBERS AND LIOUVILLE'S THEOREM

3.1. Algebraic Numbers

Attempts to solve Diophantine equations by factoring them in some extension field of $\underset{\sim}{Q}$ may date back to antiquity (see, for example, Dickson [28], p. 341). This was probably the method that led Fermat to his famous marginal assertion (Dickson [28], p. 731), and it was definitely used by Euler (Dickson [28], pp. 545 and 546). However, field extensions were not adequately understood until the nineteenth century, when Kronecker developed the algebraic theory of field extensions (roughly, Chapter 5 of van der Waerden [123]) and Gauss published the "fundamental theorem of algebra":

Theorem 3.1.1. Let $P = P(x) \in \underset{\sim}{C}[x]$ and $d(P) = n$. Then there are $a;\alpha_1,\ldots,\alpha_n \in \underset{\sim}{C}$ such that

$$P(x) = a(x - \alpha_1)\cdots(x - \alpha_n). \tag{3.1.1}$$

For proofs, see van der Waerden [123], pp. 223 and 224.

Definition 3.1.1. We call α an algebraic number if it is the root of a polynomial $P(x) = a_n x^n + a_{n-1}x^{n-1} + \cdots + a_0$, where $a_i \in \underset{\sim}{Z}$, $i = 0,\ldots,n$. The algebraic number α is said to be an algebraic integer if $a_n = 1$. The (uniquely determined up to a constant) polynomial in $\underset{\sim}{Q}[x]$ of minimal degree that α satisfies is called the minimal polynomial of α. The roots of the minimal polynomial of α are generally written as $\alpha^{(1)},\ldots,\alpha^{(n)}$ [with $\alpha = \alpha^{(1)}$] and are called the <u>conjugates</u> of α; clearly any one conjugate determines the set of all n.

From now on we fix as the minimal polynomial of an algebraic α that polynomial $P(x) = a_n x^n + \cdots + a_0$ for which $a_i \in \underset{\sim}{Z}$, $0 \le i \le n$, $a_n > 0$, and $(a_0,\ldots,a_n) = 1$. [Some authors normalize $P(x)$ so that $a_n = 1$; in that case an algebraic α is an algebraic integer if and only if the coefficients of its minimal polynomial are integers.]

<u>Definition 3.1.2</u>.

(i) Let $P = P(x) = a_n x^n + \cdots + a_0$ be any polynomial with complex coefficients. We call n its <u>degree</u> and write $d(P) = n$. The number a_n is called the <u>lead coefficient</u> of P. The quantity $H(P) = \max_{i=0}^{n} |a_i|$ is called the <u>height</u> of P, and $L(P) = |a_n| + \cdots + |a_0|$ is called the <u>length</u> of P. We define $L_\mu(P)$, the <u>μ-length</u> of P, by $L_\mu^\mu(P) = |a_n|^\mu + \cdots + |a_0|^\mu$. Thus $L(P) = L_1(P)$, and $H(P) = L_\infty(P)$. These definitions extend in the obvious way to polynomials in several variables.

(ii) If α is an algebraic number, let $P = P(x) \in \underset{\sim}{Z}[x]$ be its minimal polynomial. Its lead coefficient is defined to be the lead coefficient of P. Similarly we define $d(\alpha)$, $H(\alpha)$, $L(\alpha)$, and $L_\mu(\alpha)$, the <u>degree</u>, <u>height</u>, <u>length</u>, and <u>μ-length</u> of α, to be the corresponding entities of P; thus $d(\alpha) = d(P)$, etc.

It is an easy consequence of the symmetric function theorem (see, for example, van der Waerden [123], pp. 78-80) that the set of all algebraic numbers forms a field. We denote this field by $\underset{\sim}{A}$.

Definition 3.1.3. A complex number that is not algebraic is said to be transcendental.

We now show that transcendental numbers exist, that is, $\underset{\sim}{C} - \underset{\sim}{A}$ is nonempty.

Theorem 3.1.2. The algebraic numbers are countable.

Proof. We denote by $|T|$ the cardinality of the set T. Let S(H,n) denote the set of algebraic numbers with height H and degree n. Clearly $|S(H,n)| \leq n(2H + 1)^{n+1}$, and hence

$$\underset{\sim}{A} = \bigcup_{H=1}^{\infty} \bigcup_{n=1}^{\infty} S(H,n). \qquad \square$$

Corollary 3.1.1 (Cantor). Since the reals are uncountable, there exist transcendental numbers.

Exercises

3.1.1. Show that an irreducible polynomial with a rational root is of degree 1.

3.1.2. Let $n,m,p \in \underset{\sim}{Z}^+$, where p is prime and $(m,p) = 1$. Show that the minimal polynomial of $\exp(2\pi im/p^n)$ is

$$\frac{x^{p^n} - 1}{x^{p^{n-1}} - 1}.$$

(Hint: Set $x = y + 1$ and apply Eisenstein's criterion.)

3.1.3. Show that Theorem 3.1.1 continues to hold with $\underset{\sim}{C}$ replaced by $\underset{\sim}{A}$ (i.e., $\underset{\sim}{A}$ is algebraically closed).

3.1.4. Let $P(x) \in \underset{\sim}{Q}[x]$ be of degree n. Show that P(x) is irreducible if and only if $P^*(x) = x^n P(1/x)$ is irreducible.

(Hint: If $P(x) = R(x)S(x)$, where $R(x), S(x) \in \mathbb{Q}[x]$ are of degrees r and s, respectively, then $y^n P(1/y) = y^r R(1/y) y^s S(1/y)$. Moreover, $P^{**}(x) = P(x)$.)

3.1.5. Show that (in the notation of Definition 3.1.2) $a_n \alpha$ is an algebraic integer. (Hint: Consider $Q(z) = a_n^{n-1} P(x)$, where $z = a_n x$.)

3.1.6. (a) If $p, q \in \mathbb{Z}$ and $n = d(\alpha)$, show that $H(p\alpha/q) \leq \max(|p|, |q|)^n H(\alpha)$.

(b) Estimate $H(i\alpha)$ in terms of $H(\alpha)$.

3.1.7. Show for $P, Q \in \mathbb{C}[x]$ that $L(P + Q) \leq L(P) + L(Q)$ and $L(PQ) \leq L(P)L(Q)$.

3.2 Liouville's Theorem

All of the main results in this volume have as their source Liouville's famous theorem of 1844. After some preliminaries we shall give three proofs of it.

Definition 3.2.1. Let α be algebraic. We say that a function $A(\alpha)$ depends effectively on α (or is recursive, effective, or effectively computable) if there is an algorithm for calculating $A(\alpha)$ from $d(\alpha)$, the degree of α, and $a_0, \ldots, a_n$, the coefficients of its minimal polynomial. We say that $A(\alpha)$ is explicitly bounded above (below) if we can actually write down an expression $c(\alpha) > 0$ involving only rational, exponential, and logarithmic functions of $d(\alpha)$ and $a_0, \ldots, a_n$ such that $|A(\alpha)| \leq c(\alpha)$ ($\geq c(\alpha)$).

For our purposes we can assume that $A(\alpha)$ has its values in $\mathbb{Q}$. The definition of "explicitly bounded" is certainly artificial, but in Section 11.1 (see the discussion after Lemma 11.1.10) we shall see why we do not want to allow too wide a class of functions for $c(\alpha)$.

The absolute value of an algebraic number is explicitly bounded above and below. More precisely, we have the following lemma:

Lemma 3.2.1. If $\alpha \neq 0$ is algebraic, $n = d(\alpha)$, and $H = H(\alpha)$, then

(3.2.1) $$1/nH \leq |\alpha| \leq nH.$$

Proof. Let $P(x) = a_n x^n + \cdots + a_0$ be the minimal polynomial of α. The upper bound is immediate unless $|\alpha| > 1$, in which case

$$a_n\alpha = -(a_{n-1} + a_{n-2}\alpha^{-1} + \cdots + a_0\alpha^{n-1})$$

implies that $|\alpha| \leq |a_n\alpha| \leq nH$. For the lower bound the same argument applies to α^{-1} [which by Exercise 3.1.4 has the minimal polynomial $\pm x^n P(1/x)$]. □

Theorem 3.2.1 (Liouville). Let α be a real algebraic number with $d(\alpha) = n \geq 2$. If $p,q \in \mathbb{Z}$ and $q > 0$, then

(3.2.2) $$|\alpha - p/q| > A(\alpha)/q^n$$

for some $A(\alpha) > 0$ that can be explicitly bounded below.

Proof. Let $P(x) = a_n x^n + \cdots + a_0$ be the minimal polynomial of α and let $H = H(\alpha)$. For real x such that $P(x) \neq 0$, the mean-value theorem yields

(3.2.3) $$|x - \alpha| = \frac{|P(x) - P(\alpha)|}{|P'(x_0)|} = \frac{|P(x)|}{|P'(x_0)|}$$

for some x_0 between x and α. First assume $|p/q - \alpha| \leq 1$, so that $|p/q| \leq 1 + |\alpha|$. Multiply the right-hand side of (3.2.3) by q^n/q^n and set $x = p/q$. Since P has no rational roots (Exercise 3.1.1), $q^n P(p/q) \geq 1$ by (2.1.4). Next, $|x_0| \leq |\alpha| + |p/q| \leq 1 + 2|\alpha| \leq 1 + 2nH$ (Lemma 3.2.1), so

$$|P'(x_0)| \leq B = B(n,H)$$

follows from the triangle inequality, and hence

(3.2.4) $$\left|\frac{p}{q} - \alpha\right| \geq \frac{\min(1,B^{-1})}{q^n} . \qquad \square$$

This is the most transparent proof, but it uses the mean-value theorem and requires the reader to verify that $\min(1,B^{-1})$ is explicit in the sense of Definition 3.2.1 (see Exercise 3.2.1). Our second proof avoids the mean-value theorem, and our third makes use of Theorem 3.1.1. Both of these may be omitted on a first reading.

Before our second proof, we note that the hypotheses of Lemma 3.2.1 yield a stronger result (see Marden [72], p. 39).

Lemma 3.2.1* (Cauchy).

(3.2.1)* $$\frac{1}{(1 + M)} < |\alpha| < 1 + M,$$

where $M = \max_{i=0}^{n-1} |a_i/a_n|$.

Proof 1. The right-hand side is immediate unless $|\alpha| \geq 1$, in which case it follows from

$$0 = |P(\alpha)| \geq |a_n\alpha^n| \cdot |1 - M(|\alpha|^{-1} + \cdots + |\alpha|^{-n})|$$
$$> |a_n\alpha^n|\left(\frac{|\alpha| - 1 - M}{|\alpha| - 1}\right).$$

The left-hand side is proved as in Lemma 3.2.1. □

In particular, $|\alpha| < 1 + H$ if $|a_n| \geq 1$.

Proof 2. Use the identity

$$(3.2.5) \qquad |x - \alpha| = \frac{|P(x)|}{|[P(x) - P(\alpha)]/(x - \alpha)|}$$

instead of (3.2.3). Fix $\varepsilon > 0$ and first assume that $|x - \alpha| \leq 2\varepsilon(2 + \varepsilon)^{-1}$, so $|x| < 2\varepsilon(2 + \varepsilon)^{-1} + |\alpha|$. Let i, r, and s denote nonnegative integers, with $r + s = i - 1$.

(i) $|\alpha| \geq 1$.

From

$$(3.2.6) \qquad \left|\frac{x^i - \alpha^i}{x - \alpha}\right| \leq i \max_{r,s}\{|\alpha|^s \max(1, |x|^r)\}$$
$$\leq i[2\varepsilon(2 + \varepsilon)^{-1} + |\alpha|]^{i-1}$$

it follows that

$$(3.2.7) \qquad \left|\frac{P(x) - P(\alpha)}{x - \alpha}\right| \leq \sum_{i=1}^{n} |a_i| i[2\varepsilon(2 + \varepsilon)^{-1} + |\alpha|]^{i-1}$$
$$\leq \frac{1}{2} n(n + 1)H[2\varepsilon(2 + \varepsilon)^{-1} + |\alpha|]^{n-1}$$
$$\leq \frac{3}{4} n^2H[1 + 2\varepsilon(2 + \varepsilon)^{-1} + H]^{n-1}$$

by the triangle inequality, Lemma 3.2.1*, and $n \geq 2$. Now set $x = p/q$ and proceed as in Proof 1. From (3.2.5) and (3.2.7) it follows that

$$(3.2.8)\qquad |p/q - \alpha| \geq 4\cdot 3^{-1}\cdot q^{-n}n^{-2}H^{-1}[1 + 2\epsilon(2 + \epsilon)^{-1} + H]^{-n+1}.$$

(ii) $|\alpha| < 1$.

Here (3.2.6) still holds unless $2\epsilon(2 + \epsilon)^{-1} + |\alpha| < 1$. In that case $|(x^i - \alpha^i)/(x - \alpha)| < i$, and (3.2.8) follows from the sharper inequality

$$(3.2.9)\qquad |p/q - \alpha| \geq 4\cdot 3^{-1}q^{-n}n^{-2}H^{-1}.$$

We can now assert Liouville's theorem in the form

$$(3.2.10)\qquad |p/q - \alpha| > 8\epsilon q^{-n}n^{-2}(2 + \epsilon)^{-n+1}H^{-n},$$

$$\text{provided}\quad 0 < \epsilon \leq \tfrac{1}{6}.$$

For the right-hand side is at most $2\epsilon(2 + \epsilon)^{-1}$, so (3.2.10) is immediate unless $|p/q - \alpha| \leq 2\epsilon(2 + \epsilon)^{-1}$. In that case it follows from (3.2.8) since

$$[1 + 2\epsilon(2 + \epsilon)^{-1} + H]^{-n+1} > 6\epsilon(2 + \epsilon)^{-n+1}H^{-n+1}$$

for $0 \leq \epsilon \leq 1/6$. □

<u>Proof 3</u>. If $\alpha = \alpha^{(1)},\alpha^{(2)},\ldots,\alpha^{(n)}$ are the roots of P, then

$$(3.2.11)\qquad |p/q - \alpha| = \frac{|q^nP(p/q)|}{|a_n|q^n}\prod_{i=2}^{n}\left|\frac{p}{q} - \alpha^{(i)}\right|^{-1}$$

by Theorem 3.1.1, and again $|q^nP(p/q)| \geq 1$. It suffices to show that

$$(3.2.12)\qquad \left|\frac{p}{q} - \alpha\right| > q^{-n}4\epsilon(2 + \epsilon)^{-n+2}(H + 1)^{-n+1},$$

$$\text{provided}\quad 0 < \epsilon \leq \tfrac{1}{9}.$$

This is immediate if $|p/q - \alpha| > 2\epsilon$. Otherwise

$$(3.2.13) \quad \prod_{i=2}^{n} \left|\frac{p}{q} - \alpha + \alpha - \alpha^{(i)}\right| \le [2\epsilon + 2(H + 1)]^{n-1}$$

$$\le (2 + \epsilon)^{n-1}(H + 1)^{n-1}.$$

Hence, by (3.2.11),

$$(3.2.14) \quad |p/q - \alpha| \ge q^{-n}(2 + \epsilon)^{-n+1}(H + 1)^{-n+1},$$

and the result follows from $(2 + \epsilon)^{-1} > 4\epsilon$, $0 \le \epsilon \le 1/9$. □

Note that the hypothesis that α is real is <u>not</u> needed in Proofs 2 or 3. For α strictly complex, it can be shown (Exercise 3.2.2) that

$$(3.2.15) \quad \left|\frac{p}{q} - \alpha\right| \ge \left|\frac{(\alpha - \bar{\alpha})}{2i}\right| \ge \frac{1}{4}\, 3^{-n^2}(n + 1)^{-n}H^{-2n}.$$

Of course, the right-hand side of (3.2.15) does not depend on q.

Although Liouville's theorem (Theorem 3.2.1) is weaker than (1.5) and hence has no direct application to Diophantine equations, it does allow us to show that specific numbers are transcendental.

<u>Corollary 3.2.1</u>.

$$\xi = \sum_{i=0}^{\infty} 10^{-i!} \text{ is transcendental.}$$

<u>Proof</u>. Clearly it is irrational. Say $d(\xi) = n \ge 2$. Let

$$\frac{p_m}{q_m} = \sum_{i=0}^{m} 10^{-i!},$$

where $p_m, q_m \in \underset{\sim}{Z}^+$ and $q_m = 10^{m!}$. Then

$$0 < A(\xi)10^{-n\cdot m!} < |p_m/q_m - \xi| = 10^{-(m+1)!} + \cdots$$
$$< 2\cdot 10^{-(m+1)!}.$$

For large m this is a contradiction. □

Numbers ξ that can be shown to be transcendental by this sort of argument are called Liouville numbers (for a precise definition see Schneider [102], pp. 2 and 3). Note that the preceding proof did not require $A(\xi)$ to be effective or explicitly bounded below.

For generalizations of Liouville's theorem, see Exercise 4.2.10 and Corollary 11.1.2. Liouville's ideas can be used to construct a set of algebraically independent numbers having the cardinality of the continuum (see Schmidt [98] or von Neumann [76]).

Exercises

3.2.1. Show that if $A = A(\alpha) > 0$ and $B = B(\alpha) > 0$ are explicitly bounded below, so is $\min[A(\alpha), B(\alpha)]$. Show that the number of real roots of $P \in \underset{\sim}{Z}[x]$ is effectively computable. (Hint: $AB/(A+1)(B+1) < A$. The second part is Sturm's theorem.)

3.2.2. Prove (3.2.15). (Hint: Use Lemma 4.1.5.)

3.2.3. Show that for $\alpha \in \underset{\sim}{A}$ there is an $A(\alpha)$, explicitly bounded below, such that

$$|p/q - \alpha| \geq A(\alpha)/q^n,$$

where $n = d(\alpha)$. (Hint: Combine the first proof of Theorem 3.2.1 with Exercises 3.2.1 and 3.2.2.)

3.2.4. Say $\alpha, \beta \in \underset{\sim}{A}$ with $d(\alpha) = n$, $d(\beta) = m$, and $\alpha \neq \beta$. Show that

$$|\beta - \alpha| > C^n/H^m,$$

where $C > 0$ is independent of α and $H = H(\alpha)$. (Hint: See Brauer [14], pp. 75-78, or the hint to Exercise 4.2.10.)

3.2.5. Consider the assertion: "If α is irrational, there are infinitely many p/q such that $|p/q - \alpha| < 1/Cq^2$." Show that this is true for $C \leq \sqrt{5}$, but false for $C > \sqrt{5}$. (Hint: This is Hurwitz's theorem. Use continued fractions or see Hardy and Wright [44], pp. 163-165.)

3.2.6. Find constants C_1, C_2 so that

(a) $|\alpha - p/q| > C_1/q^2H$ if $d(\alpha) = 2$ and $H(\alpha) = H$.

(b) $|\alpha - p/q| > C_2/q^3H^2$ if $d(\alpha) = 3$ and $H(\alpha) = H$.

How large can C_1, C_2 be? Can H, H^2 be replaced by smaller powers of H?

*3.2.7. Do there exist absolute constants $C_1, C_2 > 0$ such that for every real algebraic α with $d(\alpha) = n$ and $H(\alpha) = H$,

$$|\alpha - p/q| \geq q^{-n}H^{-1} \min[C_1 4^{-n}, C_2(H + 1)^{-n+2}]?$$

3.2.8. With the hypothesis of Exercise 3.2.7 show that $|\alpha - p/q| < t < 1$ implies

$$|\alpha - p/q| \geq q^{-n}H^{-1} \min[12^{-1} \cdot 8^{-n}, (32)^{-1}(H + 1 + 4t)^{-n+2}].$$

(Hint: In Exercise 3.2.9 choose $\epsilon = 1/4$, $\gamma = 1/4$, $\theta = 1/2$, so that $\epsilon_1 = 1/24$, $\epsilon_2 = 3$. Take $\eta < 4$ and set $t = \eta/4$.)

3.2.9. Let α be a real algebraic number, $d(\alpha) = n \geq 2$, and $H(\alpha) = H$. Let $0 < \epsilon < 1/2$, $\epsilon_1 = (1/2 - \epsilon)^2/(2 - 2\epsilon)$, $\epsilon_2 = (2 - 2\epsilon)/(1 - 2\epsilon)$, $0 < \gamma < 1$, $0 < \theta < 1$, and $0 < \eta < (1 - \theta)/\gamma\theta(1 - 2\epsilon)$. Show that

(a) $|\alpha - p/q| \geq \epsilon^2 q^{-n} H^{-1}$ for $|\alpha| < 1 - 2\epsilon$.

(b) $|\alpha - p/q| \geq \min[(1 + 2\epsilon)^{-1}, \gamma\eta, \epsilon^2\theta q^{-n} H^{-1}(H + 1 + \eta)^{-n+2}]$

for $(1 - 2\epsilon)^{-1} < |\alpha|$.

(c) $|\alpha - p/q| \geq \min[2 - 2\epsilon, \gamma\eta, \epsilon_1 q^{-n} H^{-1} 2^{-n-1}(\epsilon_2 + \gamma\eta)^{-n+1}]$

otherwise.

(Hint: Without loss of generality, assume that $\alpha > 0$. Let $P(x)$ be the minimal polynomial of α.

(a) If $|x - \alpha| < \epsilon$, then $x < 1 - \epsilon$, so $|(x^i - \alpha^i)/(x - \alpha)| < i(1 - \epsilon)^{i-1}$ and $|[P(x) - P(\alpha)]/(x - \alpha)| < \Sigma_{i=0}^{n} |a_i| i(1 - \epsilon)^{i-1} < H/\epsilon^2$. Next, let $Tz = (az + b)/(cz + d)$, $A = ap + bq$, and $B = cp + dq$, where $a,b,c,d \in \underset{\sim}{Z}$ and $ad - bc = D$. Then if $p/q \neq -d/c$ and $|T\alpha| < 1 - 2\epsilon$,

$$\left|\frac{a\alpha + b}{c\alpha + d} - \frac{A}{B}\right| \geq \frac{\epsilon^2}{B^n H(T\alpha)}.$$

Hence

$$|\alpha - p/q| \geq \frac{\epsilon^2 |c\alpha + d|}{B^{n-1} q |D| H(T\alpha)}.$$

Now assume that $|\alpha - p/q| < \gamma\eta$, so $|p| < q(|\alpha| + \gamma\eta)$ and $|B| < q(|c||\alpha| + |d| + |c|\gamma\eta)$. Hence for all $p/q \neq -d/c$,

$$|\alpha - p/q| \geq \frac{\epsilon^2 |c\alpha + d|}{q^n H(T\alpha) |D| (|c||\alpha| + |d| + |c|\gamma\eta)^{n-1}}.$$

This can be stated in an ostensibly more general form: Let m be a fixed integer such that

(i) $|T(\alpha + m)| < 1 - 2\epsilon$,

(ii) $|\alpha - p/q| < \gamma\eta$,

(iii) $m + p/q \neq -d/c$.

Then $|\alpha + m - (p + mq)/q| < \gamma$ and $(p + mq)/q \neq -d/c$, so

$$(*) \quad |\alpha - p/q| \geq \frac{\epsilon^2 |c(\alpha + m) + d|}{q^n H(T(\alpha + m)) |D| (|c| |\alpha + m| + |d| + |c| \gamma\eta)^{n-1}}.$$

For (b), let a = 0, b = c = 1, d = 0, and m = 0. Then by (*),

$$|\alpha - p/q| \geq \epsilon^2 |\alpha| q^{-n} H^{-1}(\alpha)(|\alpha| + \gamma\eta)^{-n+1}$$

$$> \epsilon^2 \theta q^{-n} H^{-1}(\alpha)(|\alpha| + \eta)^{-n+2},$$

provided that $|\alpha - p/q| < \gamma\eta$ and $p/q \neq 0$.

For (c), let a = 0, b = c = 1, d = 0, and m = 1. Then $T(\alpha + 1) \leq 1 - 2(1/2 - \epsilon)/(2 - 2\epsilon)$, so by (*)

$$|\alpha - p/q| \geq \epsilon_1 q^{-n} H^{-1}(\alpha + 1)(\epsilon_2 + \gamma\eta)^{-n+1},$$

provided that $|\alpha - p/q| < \gamma\eta$ and $1 + p/q \neq 0$. Estimate $H(\alpha + 1)$ as in Exercise 3.2.12.)

In view of the results of the forthcoming chapters it might seem that Liouville's theorem and its titivations are no longer of interest. However, a remark of Baker's (see Baker [2], p. 376, footnote) shows that the subject may still have some life in it.

<u>3.2.10</u>. Is Lemma (3.2.1) best possible?

3.2.11. Obtain Lemma (3.2.1)* by showing that

$$|\alpha| < \{1 + [\sum_{j=0}^{n-1} |a_j|^p/|a_n|^p]^{q/p}\}^{1/q} \le (1 + n^{q/p}M^q)^{1/q}$$

for $p,q > 1$ such that $1/p + 1/q = 1$, and taking the limit of the right-hand side as $p \to \infty$. (Hint: Use Hölder's inequality (Marden [72], pp. 124 and 125).)

3.2.12. Estimate $H(\alpha + p/q)$, $p,q \in \mathbb{Z}$, in terms of $H(\alpha)$. (Hint: $H(\alpha + p/q) \le q^n \max_{0\le k\le n} \binom{n+1}{k} |p/q|^k H(\alpha)$, and this is largest when k is near $|p/q|(n + 1)(|p/q| + 1)^{-1}$. In particular,
$H(\alpha + 1) \le \binom{n+1}{[\frac{n+1}{2}]} H(\alpha)$.)

3.2.13. If $a,b,c,d \in \mathbb{Z}$ and $ad - bc \ne 0$, estimate $H[(a\alpha + b)/(c\alpha + d)]$ in terms of $H(\alpha)$. (Hint: If $c \ne 0$, set $D = ad - bc$, $a/c = a_1/c_1$, $D/c = D_1/c_2$, and $d/c = d_1/c_3$, where $(a_1,c_1) = (D_1,c_2) = (d_1,c_3) = 1$. By Exercises 3.1.6 and 3.2.12,

$$H\left(\frac{a\alpha + b}{c\alpha + d}\right) = H\left[\frac{a}{c} - \frac{D}{c(c\alpha + d)}\right]$$

$$\le |c_1|^n \max \binom{n+1}{k} \left(\frac{a}{c}\right)^k H\left[\frac{D}{c(c\alpha + d)}\right]$$

and

$$H\left[\frac{D}{c(c\alpha + d)}\right] \le \max(|D_1|, |c_2|)^n H(c\alpha + d)$$

$$\le \max(|D_1|, |c_2|)^n |c|^n |c_3|^n \max \binom{n+1}{k} \left(\frac{d}{c}\right)^k H(\alpha).$$

Perhaps one can do better.)

For α a real algebraic number, let $\{\alpha\}$ denote the fractional part of α (i.e., $0 < \{\alpha\} < 1$, $\alpha - \{\alpha\} \in \mathbb{Z}$).

3.2.14. Show that $d(\alpha) = d(\{\alpha\})$ and $H(\{\alpha\}) \le [2H(\alpha) + 4]^{n+1}$. (Hint: Use Exercise 3.2.12.)

3.2.15. Show that $\{\alpha\} \ge 1/(1 + [2H(\alpha) + 4]^{n+1})$ if $d(\alpha) \ge 2$.

3.2.16. Estimate $\{n\alpha\}$ and $\{\alpha^n\}$, $n \in \underset{\sim}{Z}^+$, in terms of $d = d(\alpha)$ and $H = H(\alpha)$. (Hint: Use Exercises 3.2.14 and 3.2.15, and the reasoning of Lemma 4.1.4.)

Let $\underset{\sim}{A}_n(H)$ denote the set of algebraic numbers of degree at most n and height at most H.

*3.2.17. Does $\alpha \in \underset{\sim}{A}_n(H)$ imply that $\alpha^{1/m} \in \underset{\sim}{A}_{mn}(H)$?

*3.2.18. Show that $|\underset{\sim}{A}_n(H)| \sim cH^{n+1}$ for some $c > 0$ (W. Schmidt).

**3.2.19. In $\underset{\sim}{A}_n(H)$ is there a subset of at least c_1H^{n+1} elements, $c_1 > 0$, such that the distance between any two of them is at least H^{-n-1}? (W. Schmidt).

3.2.20. Show that $\Sigma_{n=0}^{\infty} 2^{-2^{f(n)}}$, where $f(n) = 2^n$, is transcendental.

3.2.21. Assume Roth's theorem (Theorem 6.1.1). Show that $\Sigma_{n=0}^{\infty} 2^{-[\theta^n]}$ is transcendental if $\theta > 2$. Here $[x]$ denotes the greatest integer in x.

3.2.22. Let $\Lambda(P) = 2^{d(P)}L(P)$. Show that α is transcendental if and only if there are distinct polynomials $P_n(x) \in \underset{\sim}{Z}[x]$ and a sequence of positive numbers c_n with $c_n \to \infty$ such that

$$0 < |P_n(\alpha)| \le \Lambda(P_n)^{-c_n} \qquad (n = 1,2,\ldots)$$

(Mahler [59], p. 249).

**<u>3.2.23</u>. If $S \subset [0,1]$ is the Cantor set ("middle thirds removed"), is $S \cap \underset{\sim}{A} \subset \underset{\sim}{Q}$? (Mahler [70], p. 103).

Chapter 4

LEMMAS ON POLYNOMIALS AND ALGEBRAIC NUMBERS

4.1. Bounds for Roots and Coefficients

This chapter describes the basic properties of polynomials with integral, real, and complex coefficients.

Theorem 4.1.1 (Gauss's lemma). For $P, Q \in \underset{\sim}{Z}[x_1, \ldots, x_n]$ let $\gamma(P)$ denote the greatest common divisor of the coefficients of P. Then

$$\gamma(PQ) = \gamma(P)\gamma(Q).$$

As a consequence, if P can be factored over $\underset{\sim}{Q}$, it can also be factored over $\underset{\sim}{Z}$. For a proof see, for example, Cassels [18], pp. 161 and 162.

It is an easy consequence of Theorem 4.1.1 with n = 1 and the symmetric function theorem that the algebraic integers form an integral domain. (One shows first that if α is the root of a monic polynomial whose coefficients are algebraic integers, then α is an algebraic integer.) Hence an algebraic integer that is rational is a rational integer.

Lemma 4.1.1. Let $\alpha \in \underset{\sim}{A}$ have the minimal polynomial $P(x) = a_n x^n + \cdots + a_0$. Let $\alpha_1, \ldots, \alpha_r$ be any r distinct conjugates of α, where $0 \leq r \leq n$. Then

$$\gamma = a_n \alpha_1 \cdots \alpha_r$$

is an algebraic integer.

Proof. It will first be shown that if the coefficients of $Q(x) = \beta_n x^n + \cdots + \beta_0$ are algebraic integers and $Q(\alpha) = 0$, then the coefficients of $Q(x)/(x - \alpha)$ are algebraic integers. This is true if $n = d(Q) = 1$. Assume that this is true for $n - 1$. Then the coefficients of

$$\frac{Q(x) - \beta_n x^{n-1}(x - \alpha)}{x - \alpha}$$

[hence also of $Q(x)/(x - \alpha)$] are algebraic integers, and the result follows by induction. Thus the coefficients (in particular, the constant term) of

$$P(x)/(x - \alpha_{r+1}) \cdots (x - \alpha_n)$$

are algebraic integers. □

Lemma 4.1.2. Let $P(x) = a_n x^n + \cdots + a_0 \in \underset{\sim}{C}[x]$. If $0 \leq m \leq n$ and $\zeta^{n+1} = 1$, then

$$\sum_{i=0}^{n} \zeta^{-im} P(\zeta^i) = (n + 1)a_m.$$

Hence if $|P(x)| \leq B$ for $x = \zeta^i$, $0 \leq i \leq n$, it follows that $|a_m| \leq B$ for $0 \leq m \leq n$.

Proof. Both sides are $\sum_{j=0}^{n} a_j \sum_{i=0}^{n} \zeta^{i(j-m)}$. □

Of fundamental importance is the following theorem:

Theorem 4.1.2. Let $P,Q \in \underset{\sim}{C}[x]$, where $d(P) = n$, $d(Q) = m$, and $P(x) = a_n x^n + \cdots + a_0$.

(i) If $\alpha_1,\ldots,\alpha_r$ $(1 \le r \le n)$ are all roots of P satisfying $1 \le |\alpha_i|$, then

$$c_1(n)L(P) \le |a_n\alpha_1\cdots\alpha_r| \le c_2(n)L(P). \tag{4.1.1}$$

(ii) It is always true that

$$c_3(n)c_3(m)L(P)L(Q) \le L(PQ) \le L(P)L(Q). \tag{4.1.2}$$

Note: The right-hand side of (4.1.2) is immediate from the triangle inequality, and the left-hand side of (4.1.1) is quite easy (see Exercise 4.2.3). The left-hand side of (4.1.2) is immediate from (4.1.1), so it suffices to show the right-hand side of (4.1.1).

Proof. Let $\gamma_1,\ldots,\gamma_t;\ \beta_1,\ldots,\beta_s;\ \alpha_1,\ldots,\alpha_r$ $(r + s + t = n)$ denote the roots of P(x) whose absolute values lie in $[0, 1/2]$, $(1/2, 1]$, and $(1,\infty)$, respectively. For $x_0 \in \mathbb{C}$ with $|x_0| = 1$,

$$2^{-s}\,\Pi|x_0 - \alpha_i| \le \Pi|x_0 - \beta_i|\Pi|x_0 - \alpha_i|$$

$$\le \frac{|P(x_0)|}{|a_n|}\,\frac{1}{\Pi|x_0 - \gamma_i|} \le \frac{2^tL(P)}{|a_n|}.$$

It easily follows from Lemma 4.1.2 that every coefficient (in particular, the constant term) of $\Pi(x - \alpha_i)$ is bounded by $2^nL(P)/|a_n|$. □

Results of this type [with a factor $c_4(n)c_4(m)$ on the right-hand side of (4.1.2)] hold with any or all of the L(P) replaced by $L_\mu(P)$, $1 \le \mu \le \infty$, because

$$L(P)/(n+1) \le H(P) \le L_\nu(P) \le L_\mu(P) \le L(P) \quad (1 \le \mu \le \nu \le \infty)$$

by Jensen's inequality (see Hardy, Littlewood, and Pólya [45], p. 28, Theorem 19). In particular, the inequalities

$$|a_n\alpha_1\cdots\alpha_r| \le L_2(P) \tag{4.1.3}$$

$$2^{-d(PQ)}L(P)L(Q) \le L_2(PQ) \tag{4.1.4}$$

obtained in 4.2 will be used henceforth, although for many purposes the values of the constants $c_i(n)$ do not matter. It is worth mentioning that "noncancellation of polynomials" [i.e., the left-hand side of (4.1.2)] does not require the fundamental theorem of algebra for its proof.

Lemma 4.1.3. If $d(P) + 1 \le |S|$, where $S \subset \{0,1,\ldots,N\}$, then there is a function $c = c(N) > 0$ such that

$$c(N)H(P) \le \max_{k\in S}|P(k)|.$$

Proof. Let $S \supseteq \{q_0,\ldots,q_n\}$, where the q_i are distinct. We can solve the system $a_nq_j^n + \cdots + a_0 = P(q_j)$, $0 \le j \le n$, for the a_i by Cramer's rule since the determinant is Vandermondian. Hence $H(P) = \max|a_i| \le \max_{k\in S}|P(k)|c_1(N)$. □

Now let $S = \{0,1,\ldots,n+m\}$, $A = \{k \in S|c(n+m)H(P) \le |P(k)|\}$, and $B = \{k \in S|c(n+m)H(Q) \le |Q(k)|\}$. By Lemma 4.1.3 $|A| \ge m+1$ and $|B| \ge n+1$, so there is a $k \in A \cap B$. Hence

$$c^2(n+m)H(P)H(Q) \le |PQ(k)| \le H(PQ)\ [(n+m)^{n+m} + \cdots + 1],$$

the desired result.

The next result is a "noncancellation law" for polynomials under composition. It may be omitted on a first reading.

Proposition 4.1.1. Let $P,Q \in \underset{\sim}{C}[z]$ be polynomials of degrees n and m, respectively, where $m \geq 1$ and Q is monic. Let $R(z) = P[Q(z)]$ be their composition. Then

$$c(n)[L(Q) + 1]^{-nm}L(P) \leq L(R) \leq L^n(Q)L(P),$$

where $c(n) > 0$ depends only on n.

Proof. The right-hand inequality follows from trivial estimates. For the left-hand inequality choose $z_0, \ldots, z_n \in \underset{\sim}{C}$ so that the $Q(z_i)$ are the distinct $(n + 1)$st roots of unity. By Lemma 3.2.1* this can be done with $|z_i| \leq L(Q) + 1$. Let $R_i = P[Q(z_i)] = a_nQ^n(z_i) + a_{n-1}Q^{n-1}(z_i) + \cdots + a_0$, $(0 \leq i \leq n)$. By Cramer's rule,

$$a_j = (\det M)^{-1} \sum_{k=0}^{n} R_kM_k,$$

where M is the matrix $Q^j(z_i)$ and the M_k are certain $n \times n$ minors of M. Since det M is Vandermondian, it cannot vanish, so

$$H(P) = \max|a_j| \leq c_1(n)(n + 1)n! \max|R_k|.$$

But $|R_k| \leq L(R)[L(Q) + 1]^{nm}$, so the result follows. □

The general form of the preceding left-hand inequality cannot be improved for m = 1; if $P(z) = (z + 1)^n$ and $Q(z) = z - 1$, then $L[P(Q)] = L(Q)^{-n}L(P)$. The preceding proof can be made to give $c(n) = (n + 1)^{\frac{1}{2}(n-1)}(n + 1)!^{-1}$ (see Exercise 4.3.19).

The following results will be used in the chapters to come.

Theorem 4.1.3. If $P \in \underset{\sim}{C}[x]$, then $H(P)^m \leq L(P^m)$.

Proof. By Lemma 4.1.2, $H(P) \leq |P(x_0)|$ for some $x_0 \in \underset{\sim}{C}$ with $|x_0| = 1$. Hence $H(P)^m \leq |P^m(x_0)| \leq L(P^m)$ by the trivial estimate. □

Lemma 4.1.4. Let α be an algebraic integer, $d(\alpha) = n \leq N$. Then

$$\alpha^N = \sum_{i=0}^{n-1} a_i(N)\alpha^i \qquad [a_i(N) \in \underset{\sim}{Z}] \tag{4.1.5}$$

with $|a_i(N)| \leq (H+1)^{N-n+1} \leq (H+1)^N$ and $H = H(\alpha)$. Note that $a_i(n) = -a_i$, where $P(x) = a_n x^n + \cdots + a_0$ is the minimal polynomial of α.

Proof. Induct on N. If $n = N$, the result is immediate. For $n < N$,

$$\alpha^{N+1} = \sum_{i=0}^{n-2} a_i(N)\alpha^{i+1} + a_{n-1}(N)(-a_{n-1}\alpha^{n-1} - \cdots - a_0),$$

so $a_i(N+1) = a_{i-1}(N) - a_{n-1}(N)a_i$, and by the induction hypothesis

$$|a_i(N+1)| \leq (H+1)^{N-n+1} + (H+1)^{N-n+1}H = (H+1)^{N-n+2}.$$

□

For example, when $\alpha^2 + \alpha - 1 = 0$, this tells us that the Fibonacci numbers grow at most like a power of 2.

Lemma 4.1.5. Let $\alpha, \beta \in \underset{\sim}{A}$, $d(\alpha) = n$, and $d(\beta) = m$. Then

(i) $L_\mu(\alpha) = L_\mu(\alpha^{-1})$; $\quad L_\mu(-\beta) = L_\mu(\beta) \qquad (0 < \mu \leq \infty)$.

(ii) $L(\alpha\beta) \leq 2^{nm} L_2^m(\alpha) L_2^n(\beta)$.

(iii) $L(\alpha + \beta) \leq 3^{nm} L_2^m(\alpha) L_2^n(\beta)$.

(iv) $L(\alpha^{1/h}) \leq 2^{nh}L_2(\alpha)$.

(v) $L(\alpha^k) \leq 2^{nk}L_2(\alpha)$.

(vi) If $W(x,y) \in \underset{\sim}{Z}[x,y]$ is of degree h and k in x and y, respectively, and $W(\alpha,\beta) = 0$ but $W(x,\beta) \neq 0$, then

$$L(\alpha) \leq 2^{mh}L^m(W)L_2^k(\beta).$$

Proof. Case (i) is left as an exercise for the reader. In the remaining cases let a and b denote the lead coefficients of α and β, and let α_i and β_j, $1 \leq i \leq n$, $1 \leq j \leq m$, be their conjugates. For (ii), let $g(x) = c\Pi_\Omega(x - \alpha_i\beta_j)$ be the minimal polynomial of $\alpha\beta$, where $c \in \underset{\sim}{Z}$ and

$$\Omega \subseteq \{(i,j) \mid 1 \leq i \leq n,\ 1 \leq j \leq m\}.$$

Then by Lemma 4.1,1, $h(x) = a^m b^n \Pi_\Omega(x - \alpha_i\beta_j) \in \underset{\sim}{Z}[x]$, so $h(x) = f\cdot g(x)$, where $f \in \underset{\sim}{Z}$ by Gauss's lemma (Theorem 4.1.1). Hence $c \mid a^m b^n$, so (ii) follows from $L(\alpha\beta) = L(g(x))$, the definition of length, and (4.1.3). The proof of (iii) is similar to the proof of (ii). For (vi), let P(x) be the minimal polynomial of α. Then by Lemma 4.1.1 and Gauss's lemma,

$$b^k \prod_{i=1}^{m} W(x,\beta_i) = P(x)Q(x), \qquad (Q \in \underset{\sim}{Z}[x]).$$

By (4.1.4), (4.1.2), and (4.1.3),

$$\begin{aligned} (4.1.6) \qquad L(\alpha) \leq L(P)L(Q) &\leq 2^{mh}L_2\left(b^k \prod_{i=1}^{m} W(x,\beta_i)\right) \\ &\leq 2^{mh}b^k \prod_{i=1}^{m} L(W(x,\beta_i)) \\ &\leq 2^{mh}L^m(W(x,y))b^k \prod_{i=1}^{m} \max(1, |\beta_i|^k) \\ &\leq 2^{mh}L^m(W)L_2^k(\beta), \end{aligned}$$

the desired result. If α,β are replaced by $\gamma^{1/h},\gamma$ and $W(x,y) = x^h - y$, then (4.1.6) becomes $L(\gamma) \leq 2^{mh}L_2(P(x^h))$, where $P(x)$ is the minimal polynomial of γ, and (iv) follows (set $\gamma = \alpha$). The proof of (v) is similar. □

Lemma 4.1.7 will be the analog for algebraic number fields of the fact that there are no integers between 0 and 1.

<u>Lemma 4.1.6</u>. Let $\beta \in \underset{\sim}{K} = \underset{\sim}{Q}(\alpha)$, $d(\alpha) = n$, and $d(\beta) = m$. Let $\sigma_1,\ldots,\sigma_n$ be all isomorphisms of $\underset{\sim}{K}$ into $\underset{\sim}{C}$. Then the sequence $\sigma_1(\beta),\ldots,\sigma_n(\beta)$ consists of m distinct numbers, each repeated n/m times.

<u>Proof</u>. Let $Q(x)$ be the minimal polynomial of β and $R(x) = \Pi_{i=1}^{n}[x - \sigma_i(\beta)]$. Then

$$R(x) = aQ(x)^e S(x),$$

where $a \in \underset{\sim}{Q}$, $e \in \underset{\sim}{Z}^+$, and $(Q(x),S(x)) = 1$. Hence $S(x) \in \underset{\sim}{Q}$, and the result follows from $e = n/m$. □

<u>Lemma 4.1.7</u>. Let $\alpha_1,\ldots,\alpha_r$ be any algebraic numbers of degrees $n_1,\ldots,n_r$, respectively. Let

$$\text{(4.1.7)} \qquad \alpha = \sum_{k_1=0}^{N_1} \cdots \sum_{k_r=0}^{N_r} A(k_1,\ldots,k_r)\alpha_1^{k_1}\cdots\alpha_r^{k_r},$$

where $A(k_1,\ldots,k_r) \in \underset{\sim}{Z}$ and

$$\text{(4.1.8)} \qquad \sum_{k_1=0}^{N_1} \cdots \sum_{k_r=0}^{N_r} |A(k_1,\ldots,k_r)| \leq B.$$

Then either $\alpha = 0$ or

$$|\alpha| \geq B^{1-n} \prod_{i=1}^{r} L_2^{-nN_i/n_i}(\alpha_i),$$

where $n = d(\underset{\sim}{K}/\underset{\sim}{Q}) \leq n_1\cdots n_r$ and $\underset{\sim}{K} = \underset{\sim}{Q}(\alpha_1,\ldots,\alpha_r)$.

Proof. Let $\sigma_1,\ldots,\sigma_n$ be all isomorphisms of $\underset{\sim}{K}$ into $\underset{\sim}{C}$. Let $a_1,\ldots,a_r$ be the lead coefficients of $\alpha_1,\ldots,\alpha_r$. Say $\alpha \neq 0$. Then by Lemma 4.1.6, Lemma 4.1.1 and the symmetric function theorem

$$s = a_1^{nN_1/n_1}\cdots a_r^{nN_r/n_r} \prod_{j=1}^{n} \sigma_j(\alpha) \tag{4.1.9}$$

is a nonzero rational integer. Next,

$$|\sigma_j(\alpha)| \leq B \prod_{i=1}^{r} \max[1, |\sigma_j(\alpha_i)|]^{N_i},$$

so

$$\prod_{j=2}^{n} |\sigma_j(\alpha)| \leq B^{n-1} \prod_{i=1}^{r} \{\prod_{j=1}^{n} \max[1, |\sigma_j(\alpha_i)|]^{N_i}\} \tag{4.1.10}$$

$$\leq B^{n-1} \prod_{i=1}^{r} \{L_2^{n/n_i}(\alpha_i)/|a_i|^{n/n_i}\}^{N_i}$$

by Lemma 4.1.6 and (4.1.3). Without loss of generality we may assume that $\alpha = \sigma_1(\alpha)$, and the result then follows from (4.1.9) and (4.1.10) since $1 \leq |s|$. □

The following results are not required in the sequel and may be omitted on a first reading. They give an estimate for the length and degree of an algebraic number defined by the intersection of two algebraic curves.

Proposition 4.1.2. Let $P = P(u_1,\ldots,u_m) \in \mathbb{C}[u_1,\ldots,u_m]$, let $Q_i = Q_i(x) \in \mathbb{C}[x]$ for $1 \leq i \leq m$, and set $R(x) = P(Q_1(x),\ldots,Q_m(x))$. Then

$$L(R) \leq L(P)L^{d_1}(Q_1)\cdots L^{d_m}(Q_m),$$

where $d_i = d_{u_i}(P)$.

Proof. This follows from trivial estimates. □

Proposition 4.1.3. Let P (as in Proposition 4.1.2) be symmetric in $u_1,\ldots,u_m$ and say $d_{u_i}(P) \leq N$ for $1 \leq i \leq m$. Let $s_1,\ldots,s_m$ be the elementary symmetric functions of $u_1,\ldots,u_m$. Then

$$P(u_1,\ldots,u_m) = Q(s_1,\ldots,s_m) \in \mathbb{C}[s_1,\ldots,s_m],$$

where $d_{s_i}(Q) \leq N$ and $H(Q) \leq H(P)\cdot 3^{m(N+1)^{m+1}}$.

Proof. As in a standard proof of the symmetric function theorem (see van der Waerden [123], pp. 78-80), order the terms of P lexicographically according to their exponents. Let $P_0 = P$ and proceed inductively. If $c_j x_1^{e_1}\cdots x_m^{e_m}$ is the first term of P_j in this ordering, define

$$P_{j+1} = P_j - c_j s_1^{e_1-e_2} s_2^{e_2-e_3}\cdots s_{m-1}^{e_{m-1}-e_m} s_m^{e_m}.$$

Thus

$$|c_{j+1}| \leq H(P_{j+1}) \leq H(P_j) + H(P_j)2^{me_1} \leq H(P_j)\cdot 3^{mN}.$$

For some $j \leq (N+1)^m$ we have $P_j \equiv 0$, so $P = P_0 = Q(s_1,\ldots,s_m)$ with

$$H(Q) \leq \max|c_j| \leq H(P)\cdot 3^{m(N+1)^{m+1}}. \qquad \square$$

<u>Proposition 4.1.4</u>. Let $W_1, W_2 \in \mathbb{Z}[x,y]$ be of degree n_1, m_1 and n_2, m_2 in x and y, respectively, where $n_1, m_1, n_2, m_2 \geq 1$. Assume that W_1 and W_2 are independent in the sense that there is no algebraic function $g(x)$ such that

$$W_1(x,g(x)) \equiv W_2(x,g(x)) \equiv 0.$$

Say $W_1(\alpha,\beta) = W_2(\alpha,\beta) = 0$, where $\alpha,\beta \in \mathbb{C}$. Then α is an algebraic number with

$$d(\alpha) \leq m_1(n_1m_2 + n_2)$$

and

$$L(\alpha) \leq (m_1n_2 + 1)(m_2 + 1)^{m_1} 2^{m_1(n_1m_2+n_2)} 3^{m_1(m_2+1)^{m_1+1}} \cdot L^{m_1}(W_2)L^{m_1m_2}(W_1).$$

<u>Proof</u>. We use the notation of the preceding propositions.

Set

$$\begin{aligned} W = \prod_{i=1}^{m_1} W_2(x,u_i) &= \prod_{i=1}^{m_1} [x^{n_2}p_{n_2}(u_i) + \cdots + p_0(u_i)] \\ &= \sum_{j=0}^{m_1n_2} x^j P_j(u_1,\ldots,u_{m_1}) \\ &= \sum_{j=0}^{m_1n_2} x^j Q_j(s_1,\ldots,s_{m_1}). \end{aligned}$$

Clearly $d_{u_i}(P_j) \leq m_2$ and $L(P_j) \leq L^{m_1}(W_2)$. By Proposition 4.1.3 we have $d_{s_i}(Q_j) \leq m_2$ and

$$H(Q_j) \le H(P_j)\cdot 3^{m_1(m_2+1)^{m_1+1}}.$$

Now let the $u_i = u_i(x)$ be the roots of $W_1(x,y) = 0$, where W_1 is considered to be a polynomial in y with coefficients involving the parameter x. The s_i become polynomials in x of degree at most n_1 and height at most $L(W_1)$, while $W = W(x) \in \underset{\sim}{Z}[x]$. Clearly $W(\alpha) = 0$, but W is not identically zero. It is immediate that $d(\alpha) \le m_1n_2 + m_2m_1n_1$. Also,

$$L(W) \le \sum_{j=0}^{m_1n_2} L(R_j) \le (m_1n_2 + 1) \max L(R_j),$$

where $R_j = R_j(x) = Q_j(s_1(x),\ldots,s_m(x))$, so by Proposition 4.1.2

$$L(R_j) \le L(Q_j)L(W_1)^{m_1m_2} \le (m_2 + 1)^{m_1}H(Q_j)L(W_1)^{m_1m_2}.$$

From Gauss's lemma (Theorem 4.1.1) we know that $W(x) = p_1(x)p_2^f(x)$, where $p_1(x),p_2(x) \in \underset{\sim}{Z}[x]$ and $p_2(x)$ is the minimal polynomial of α; thus (4.1.4) yields

$$L(W) \ge 2^{-d(W)}L(p_1)L^f(p_2) \ge 2^{-m_1(n_2+m_2n_1)}L(\alpha).$$

The result follows by combining the inequalities displayed above.

$\square$

Exercises

Here Exercises 4.1.1 through 4.1.5 lead to another proof (due to Sprindžuk [117], pp. 25 and 26; also [116]) of Theorem 4.1.2(ii). Exercises 4.1.7 and 4.1.8 are also in Sprindžuk [117], pp. 26 and 27, and pp. 32 and 33.

4.1.1. If $P(x) = a_n x^n + \cdots + a_0$ and $|a_n| > cH(P)$, then $|\alpha| < n/c$ whenever $P(\alpha) = 0$.

4.1.2. Show that $\max_{m=0}^{n} |P(m)| \geq c_1(n)H(P)$. (Hint: Use the Lagrange interpolation formula.)

4.1.3. If $|\alpha| \leq B$ whenever $P(\alpha) = 0$, then $H(P) \leq c_2(n,B)|a_n|$.

4.1.4. Let $P^*(x) = x^n P(1/x)$; clearly $H(P^*) = H(P)$. Define $P_{(m)}(x) = P(x + m)$ for $m = 0,1,\ldots,n$ and $P^*_{(m)}(x) = [P_{(m)}(x)]^*$. Show that when $P(m) \neq 0$, it is the lead coefficient of $P^*_{(m)}(x)$, and

$$c_3(n)H(P) \leq H(P_{(m)}(x)) \leq c_4(n)H(P).$$

(Hint: For the lower bound use $P(x) = P((x + m) - m)$.)

4.1.5. Show that $H(P_1P_2) \geq c_5(n)H(P_1)H(P_2)$. (Hint: Set $P = P_1P_2$ and $Q = P^*_{(m)}(x) = P^*_{1(m)}(x)P^*_{2(m)}(x)$, where $|P(m)| = \max_{i=0}^{n} |P(i)|$.)

Definition. Say $Q \cong P$ (Q is equivalent to P) for polynomials Q,P if Q can be obtained from P by a finite number of operations of the form $P \to P^*$, $P \to P_{(m)}$, where $m \in \mathbb{Z}$.

4.1.6. If $a,b,c,d \in \mathbb{Z}$ and $ad - bc = \pm 1$, show that

$$P(x) \cong (cx + d)^n P\left(\frac{ax + b}{cx + d}\right),$$

where $n = d(P)$. (Hint: This is not hard if one knows that the full modular group is generated by $U = \begin{pmatrix} 1 & 1 \\ 0 & 1 \end{pmatrix}$ and $T = \begin{pmatrix} 0 & -1 \\ 1 & 0 \end{pmatrix}$. (For a very simple proof see Magnus, Karrass, and Solitar [63], p. 44, Exercise 18.) For then $U = B$ and $T = B^{-1}ABAB^{-1}$, where $A = \begin{pmatrix} 0 & 1 \\ 1 & 0 \end{pmatrix}$ and $B = \begin{pmatrix} 1 & 1 \\ 0 & 1 \end{pmatrix}$ correspond to the operations $P \to P^*$ and $P \to P_{(1)}$, respectively.)

4.1.7. Let P,Q,R be elements of $\mathbb{Z}[x]$. Show that

(a) If $P \cong Q$ and P is irreducible, then Q is irreducible.

(b) Given P, there is an $r_n x^n + \cdots + r_0 = R \cong P$ such that

(i) $H(R) = |r_n|$,

(ii) $c_1(n)H(P) < H(R) < c_2(n)H(P)$.

(Hint: Define m by $|P(m)| = \max_{i=0}^{n} |P(i)|$. Then $P \cong P^*_{(m)} = Q = q_n x^n + \cdots + q_0$, where $c_3(n)H(P) < |q_n|$ by Exercises 4.1.4 and 4.1.2, and $c_4(n)H(P) < H(Q) < c_5(n)H(P)$ by Exercise 4.1.4. Note that if the hypothesis of Exercise 4.1.1 holds for any polynomial Q, it holds also for the derivative Q'. Hence

$$|Q^{(k)}(x)| \le n!\, |q_n| \prod_{i=1}^{n-k} |x - \beta_i^{(k)}|$$

$$< |q_n| \prod_{i=1}^{n} |x - \beta_i| = |Q(x)|$$

for all $k \ge 1$, $x \ge r = r(n)$, where the β_i $[\beta_i^{(k)}]$ are the roots of Q $[Q^{(k)}]$. Now show that the constant term is the largest coefficient of $Q(x + r)$ (expand in a Taylor series) and set $R = Q^*_{(r)}(x)$.)

For $\xi_1, \xi_2 \in \mathbb{C}$ say $\xi_1 \cong \xi_1$ if $\xi_1 = (a\xi_2 + b)/(c\xi_2 + d)$, where $a,b,c,d \in \mathbb{Z}$ and $ad - bc = \pm 1$.

4.1.8. If $|P(\xi_1)| < c(n)/H(P)^{\omega}$ has infinitely many solutions $P \in Z[z]$ with $d(P) \leq n$, where $\omega > 0$ is fixed, show that for some $\xi_2 \cong \xi_1$, $|P(\xi_2)| < c(n)/H(P)^{\omega}$ has infinitely many solutions [with perhaps a different $c(n)$] $P = a_n x^n + \cdots + a_0$ such that $H(P) = |a_n|$ and P is irreducible. (Hint: Use Exercise 4.1.5 and the reasoning of Exercise 4.1.7; see also Exercise 4.2.17.)

4.2. Improved Bounds

Inequalities more precise than (4.1.1) and (4.1.2) [such as (4.1.3) and (4.1.14)] will be obtained here. This section may be omitted on a first reading.

Let D be an open subset of $\mathbb{C}$. Denote by $H(D)$ the set of functions holomorphic on D (i.e., these functions that have a derivative at every point of D). Then $f \in H(D)$ implies that $\mathrm{Re}(f)$ and $\mathrm{Im}(f)$, the real and imaginary parts of f, are harmonic on D. If $h(z)$ is real and harmonic on D, and D contains a closed disk of radius R about z_0, then

$$h(z_0) = \frac{1}{2\pi}\int_0^{2\pi} h(z_0 + Re^{i\theta})\, d\theta. \tag{4.2.1}$$

Denote the closure of a set $E \subseteq \mathbb{C}$ by $\overline{E}$.

Lemma 4.2.1 (Jensen). Let $E = \{z \mid |z| < R\}$. Assume that $f \in H(D)$, where $D \supset \overline{E}$, $f(0) \neq 0$, and $f(z) \neq 0$ if $|z| = R$. Let

$\alpha_1,\ldots,\alpha_n$ be all zeros of f inside E (m-fold zeros listed m times). Then

$$\log \prod_{j=1}^{n} \frac{R}{|\alpha_j|} = \frac{1}{2\pi}\int_0^{2\pi} \log\left|\frac{f(Re^{i\theta})}{f(0)}\right| d\theta.$$

<u>Proof</u>. Let

$$g(z) = f(z) \prod_{j=1}^{n} \frac{R^2 - \bar{\alpha}_j z}{R(z - \alpha_j)} . \tag{4.2.2}$$

Clearly g has no zeros on some open set F such that $F \supset \bar{E}$. Define $\log g(z)$ so that $\log g(z) \in H(F)$. Then

$$\log|g(z)| = \mathrm{Re}[\log g(z)]$$

is harmonic on F. From (4.2.1) it follows that

$$\log|g(0)| = \frac{1}{2\pi}\int_0^{2\pi} \log|f(Re^{i\theta})|\, d\theta$$

since each factor of the product on the right-hand side of (4.2.2) has absolute value 1 when $|z| = R$. □

Now say $P \in \underset{\sim}{C}[x]$ has no roots on the unit circle. Let $n = d(P)$ and write

$$P(x) = a_n x^n + \cdots + a_0 = a_n(x - \alpha_1)\cdots(x - \alpha_n),$$

where

$$|\alpha_1| \le \cdots \le |\alpha_r| < 1 < |\alpha_{r+1}| \le \cdots \le |\alpha_n| \qquad (0 \le r \le n)$$

[$r = 0(n)$ means that all roots are greater (less) than unity in absolute value]. By Lemma 4.2.1 with $R = 1$, we find that

$$(4.2.3)\quad \log|a_n\alpha_{r+1}\cdots\alpha_n| = \log\left|\frac{a_0}{\alpha_1\cdots\alpha_r}\right| = \frac{1}{2\pi}\int_0^{2\pi} \log|P(e^{i\theta})|\,d\theta.$$

By exponentiating (4.2.3) and applying the inequality of the arithmetic-geometric means, the Cauchy-Schwarz inequality, and Parseval's formula, we obtain

$$(4.2.4)\quad \begin{aligned} |a_n\alpha_{r+1}\cdots\alpha_n| &\le \frac{1}{2\pi}\int_0^{2\pi} |P(e^{i\theta})|\,d\theta \\ &\le \frac{1}{2\pi}\left(\int_0^{2\pi} d\theta\right)^{1/2}\left(\int_0^{2\pi} |P(e^{i\theta})|^2\,d\theta\right)^{1/2} \\ &\le L_2(P). \end{aligned}$$

Since both sides of (4.2.4) are continuous functions of the roots of P, the inequality is also valid when P has roots on the unit circle. This proves (4.1.3).

Next,

$$L(P) \le |a_n| \prod_{i=1}^{n} (1 + |\alpha_i|) \le 2^n|a_n\alpha_{r+1}\cdots\alpha_n|,$$

so for any polynomials P,Q without zeros on the unit circle, (4.2.3) yields

$$(4.2.5)\quad \begin{aligned} \log L(P) + \log L(Q) &\le [d(P) + d(Q)] \log 2 \\ &+ \frac{1}{2\pi}\int_0^{2\pi} \log|P(e^{i\theta})Q(e^{i\theta})|\,d\theta. \end{aligned}$$

Now (4.1.4) follows from (4.2.5) in the same way as (4.1.3) followed from (4.2.3). In fact this shows that if

$P(x) = \Pi_{i=1}^{m} P_i(x) \in \underset{\sim}{C}[x]$, then

$$\prod_{i=1}^{m} L(P_i) \leq 2^{d(P)} L_2(P).$$

The above estimates are due to Mahler [68] and are also discussed by Marden [72], p. 129 (see also Duncan [29] and Mahler [69]). Duncan's improvement is based on a famous identity for the sum of the squares of the binomial coefficients. Let log M(P) denote the right-hand side of (4.2.3). Then

$$L_2^2(P) = \sum_{j=0}^{n} a_i^2 \leq |a_n\alpha_{r+1}\cdots\alpha_n|^2 \sum_{j=0}^{n} \binom{n}{j}^2$$

$$= \binom{2n}{n} M^2(P) \leq \binom{2n}{n} L_2^2(P).$$

The last inequality is simply (4.2.4). Hence if $d_i = d(P_i)$, we have

$$\prod_{i=1}^{m} \binom{2d_i}{d_i}^{-1/2} L_2(P_i) \leq \prod_{i=1}^{m} M(P_i) = M\left(\prod_{i=1}^{m} P_i\right)$$

$$\leq L_2(P).$$

Note that $\binom{2n}{n}^{1/2}$ tends to infinity more slowly than 2^n.

Inequality (4.1.3) could also be deduced from the following much stronger lemma:

Lemma 4.2.2 (Gonçalves). Let

$$P(x) = a_n x^n + \cdots + a_0 = a_n(x - \alpha_1)\cdots(x - \alpha_n) \in \underset{\sim}{C}[x].$$

Then

$$|a_n\alpha_1\cdots\alpha_r|^2 + |a_n\alpha_{r+1}\cdots\alpha_n|^2 \leq L_2^2(P) \qquad (0 \leq r \leq n),$$

and this is best possible [e.g., consider $P(x) = x^n - 1$].

Proof. Let $P(x) = f(x)g(x)$, where

$$f(x) = b_r x^r + \cdots + b_0 = b_r(x - \alpha_1)\cdots(x - \alpha_r)$$

and

$$g(x) = c_s x^s + \cdots + c_0 = c_s(x - \alpha_{r+1})\cdots(x - \alpha_n),$$

so $s = n - r$. Set $h(x) = x^s\bar{g}(1/x) = \bar{c}_s + \cdots + \bar{c}_0 x^s$ and $Q(x) = f(x)h(x)$. Clearly

$$|c_0 b_r|^2 + |c_s b_0|^2 \leq L_2^2(Q). \tag{4.2.6}$$

Gonçalves [41] showed by a direct calculation involving double sums that $L_2^2(Q) = L_2^2(P)$, so the result follows on dividing (4.2.6) by $a_n^2 = b_r^2 c_s^2$. Ostrowski [81] later pointed out that since $|h(e^{i\theta})| = |g(e^{i\theta})|$,

$$L_2^2(P) = \frac{1}{2\pi}\int_0^{2\pi} |P(e^{i\theta})|^2\, d\theta = \frac{1}{2\pi}\int_0^{2\pi} |f(e^{i\theta})|^2 |h(e^{i\theta})|^2\, d\theta$$

$$= L_2^2(Q).$$

□

The factor $2^{-d(PQ)}$ in (4.1.4) is generally not best possible, and only in some special cases is a sharp result known. Another result of this type, applicable to polynomials in s variables, will now be obtained.

Lemma 4.2.3. Let $P(x) = a_n(x - \alpha_1)\cdots(x - \alpha_n)$, where $\alpha_i = |\alpha_i|\zeta_i \neq 0$, and define $\|P\| = \max_{|x|=1} |P(x)|$. If $|x| = 1$, then

UNIVERSITY LIBRARY

$$2^{-n}\|P\| \prod_{i=1}^{n} |x - \zeta_i| \le |P(x)|.$$

Proof. The point on a chord of a circle closest to the center of the circle is the midpoint of the chord. Hence

$$(1 + |\alpha_i|) \frac{|x - \zeta_i|}{2} \le (1 + |\alpha_i|) |[\frac{x}{1 + |\alpha_i|} + \frac{(-\zeta_i)|\alpha_i|}{1 + |\alpha_i|}]|$$

$$= |x - \alpha_i|,$$

and the result follows from $\|P\| \le a_n \prod_{i=1}^{n} (1 + |\alpha_i|)$. □

Theorem 4.2.1 (Gelfond). If $P(x_1,\ldots,x_s) = \prod_{i=1}^{m} P_i(x_1,\ldots,x_s)$, where $P_i \in \underset{\sim}{C}[x_1,\ldots,x_s]$, then

$$\prod_{i=1}^{m} L_2(P_i) \le 2^{d(P)-(s/2)} L_2(P).$$

Proof. Let $n_i = d_i(P)$ denote the degree of P in x_i and set $x_i = \exp(2\pi i\theta_i)$. Then Lemma 4.2.3 [with $x = x_1$ and $P(x_1) = P(x_1,\ldots,x_s)$] together with a trivial estimate yields

$$(4.2.7) \quad 2^{-2n_1} \prod_{i=1}^{n_1} |x_1 - \zeta_i|^2 \int_0^1 \cdots \int_0^1 |P_1(x_{11},x_2,\ldots,x_s)|^2 \cdots$$

$$|P_m(x_{1m},x_2,\ldots,x_s)|^2 \, d\theta_{11} \cdots d\theta_{1m} \le |P|^2,$$

where $x_{ij} = \exp(2\pi i\theta_{ij})$ and the ζ_i are roots of unity that depend on the variables $x_2,\ldots,x_s$ (temporarily held constant). Now

$$2 \leq \int_0^1 \left| \prod_{j=1}^{n_1} [\exp(2\pi i\theta_1) - \zeta_j] \right|^2 d\theta_1$$

for any positive integer n_1 (the first and last coefficients of the integrand each contribute 1), so the integration of (4.2.7) with respect to θ_1 yields

$$\text{(4.2.8)} \qquad F_1\colon\ 2^{-2n_1+1} \int_0^1 \cdots \int_0^1 |\cdots|^2\, d\theta_{11} \cdots d\theta_{1m} \leq \int_0^1 |P|^2\, d\theta_1 .$$

Let F_j denote the formula that would be obtained here if our initial variables were $x_j, x_{j+1}, \ldots, x_s$. If both sides of F_1 are integrated with respect to θ_2, and the integral on θ_2 is brought inside the first m integrals on the left-hand side of (4.2.8), we see that F_2 is applicable to this innermost integral. Next, integrate with respect to θ_3, interchange, apply F_3, etc. After F_s has been applied, we have

$$2^{-2(n_1+\cdots+n_s)+s}\, L_2^2(P_1) \cdots L_2^2(P_m) \leq \int_0^1 \cdots \int_0^1 |P|^2\, d\theta_1 \cdots d\theta_s$$
$$= L_2^2(P). \qquad \square$$

<u>Theorem 4.2.2 (Gelfond)</u>. If $P(x_1, \ldots, x_s) \in \mathbb{C}[x_1, \ldots, x_s]$ is of degree n_i in x_i, $1 \leq i \leq s$, then

$$L_2^m(P) \leq L_2(P^m) \prod_{i=1}^{s} (1 + 2mn_i)^{1/2}.$$

<u>Proof</u>. Let $\|P\|$ denote the maximum value of P on the set $|x_i| = 1$, $1 \leq i \leq s$. From the finite Fourier identity

$$|P|^{2m} = \sum_{k_1=-mn_1}^{mn_1} \cdots \sum_{k_s=-mn_s}^{mn_s} \{\int_0^1 \cdots \int_0^1 |P^m|^2 \exp[-2\pi i(k_1\theta_1 + \cdots + k_s\theta_s)]$$

$$d\theta_1 \cdots d\theta_s\} \exp[2\pi i(k_1\theta_1 + \cdots + k_s\theta_s)],$$

where $x_i = \exp(2\pi i\theta_i)$, it follows by the trivial estimate that

$$\|P\|^{2m} \leq L_2^2(P^m) \prod_{i=1}^{s} (1 + 2mn_i),$$

whereas

$$L_2^{2m}(P) = (\int_0^1 \cdots \int_0^1 |P|^2 \, d\theta_1 \cdots d\theta_s)^m \leq \|P\|^{2m}.$$

□

Note that Theorem 4.2.2 is stronger than the result obtained from Theorem 4.2.1 by equating all the P_i. But it will now be shown that Theorem 4.2.1 is "nearly best possible" when $s = 1$. For $P \in \underset{\sim}{C}[x]$, $d(P) = n$, and $P = P_1(x) \cdots P_m(x)$, set

$$\chi(P_1, \ldots, P_m; n) = \chi(n) = \log_2 \frac{L_2(P_1) \cdots L_2(P_m)}{L_2(P)} .$$

Here $\chi(n)$ measures the "amount of cancellation" that takes place in the product $P_1 \cdots P_m$. Certainly $\chi(n) \leq n - 1/2$.

<u>Theorem 4.2.3 (Gelfond)</u>. For some absolute constants $c_1, c_2 > 0$,

$$n - c_1 n^{1/3} \log n \leq \chi(n) \leq n - c_2 n^{1/3} \log n$$

holds infinitely often.

<u>Proof</u>. Let $P(x) = x^n - 1 = \Pi_{k=1}^m P_k(x)$, where $n = mh$ and

$$P_k = \prod_{\ell=0}^{h-1} (x - \exp\{\frac{2\pi i[(k-1)h + \ell]}{n}\}).$$

Here n is always chosen to be a perfect cube, $h = n^{2/3}$, and $m = n^{1/3}$. From

$$L_2^2(P_k) \le \sum_{\ell=0}^{h} \binom{h}{\ell}^2 = \binom{2h}{h}$$

and Stirling's formula it follows that

$$L_2(P_k) \le c_3 2^{h-c_4 \log h}.$$

For a lower bound set $\phi = 2\pi h/n$. From Figure 2 we see that there is an angle θ such that

$$L_2^2(P_k) = \frac{1}{2\pi}\int_0^{2\pi} |P_k(e^{i\theta})|^2 \, d\theta$$

$$\ge \frac{1}{2\pi}\int_\theta^{\theta+\phi} (2\cos\frac{\pi h}{n})^{2h} \, d\theta = n^{-1}h2^{2h}(\cos\frac{\pi h}{n})^h.$$

For n sufficiently large,

$$(\cos\frac{\pi h}{n})^h > (1 - \frac{c_5 h^2}{n^2})^h \to e^{-c_5} > 0.$$

Hence

$$c_6 2^{h-c_7 \log h} \le L_2(P_k),$$

and the result easily follows.

□

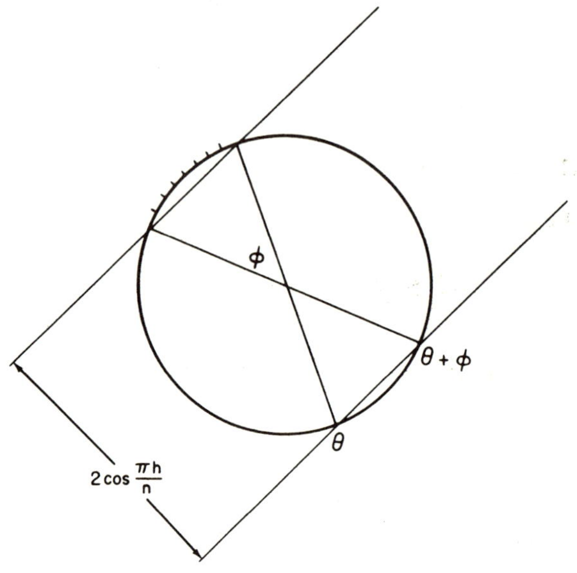

Figure 2

Now are there positive constants c, γ, $0 < \gamma < 1/3$, such that $\chi(n) \leq n - cn^{\gamma}$? This question seems to be open.

Theorems 4.2.1, 4.2.2, and 4.2.3 can be found in Gelfond [39], pp. 135-140.

Exercises

Many of the exercises listed here are results obtained by Wirsing [124]. Throughout let $P \in \underset{\sim}{Z}[x]$, $d(P) = n$, and write $\alpha = \alpha^{(1)}$, where

$$P(x) = a_n x^n + \cdots + a_0 = a_n(x - \alpha^{(1)}) \cdots (x - \alpha^{(n)}).$$

4.2.1. Show that the following are true:

(a) $H(cP) = |c|H(P)$ for $c \in \underset{\sim}{C}$, and $H(P(cx)) \leq (1,|c|^n)H(P)$.

(b) $H(x^nP(1/x)) = H(P(x)) = H(P(-x))$.

(c) $H(P^{(\nu)}(x)) \leq c(n)H(P(x))$, where $\nu \in \underset{\sim}{Z}$ and $P^{(\nu)}(x)$ denotes the ν-th derivative of $P(x)$. (Hint: $H(P^{(\nu)}(x)) \leq \nu!2^nH(P(x))$.)

4.2.2. Show that $H(P(x + \xi)) \leq c(n,\xi)H(P(x))$ for $\xi \in \underset{\sim}{C}$. (Hint: Write $P(x + \xi) = \Sigma_{\nu=0}^{n} P^{(\nu)}(\xi)x^{\nu}/\nu!$ and apply Exercise 4.2.1.)

4.2.3. Show that $2^{-n}L(P) \leq |a_n| \Pi'|\alpha^{(\nu)}| \leq L(P)$, where the prime indicates that the product is restricted to $|\alpha^{(\nu)}| \geq 1$. (Hint: The right-hand side is (4.1.3), while

$$\Sigma|a_i| \leq |a_n|\Pi(1 + |\alpha^{(\nu)}|) \leq |a_n|2^{n-s}\,\Pi'(1 + |\alpha^{(\nu)}|),$$

where s is the number of factors in Π'.)

Note that $c_1(n)\,\Pi_{i=1}^{m} L(P_i) \leq L(\Pi_{i=1}^{m} P_i) \leq c_2(n)\,\Pi_{i=1}^{m} L(P_i)$ is a consequence of Exercise 4.2.3 (Wirsing [124]). For yet another proof see Schneider [103], pp. 75-77.

4.2.4. Show that for $0 < \psi < \infty$ and $\xi \in \underset{\sim}{C}$,

$$\frac{c_1(n,\xi,\psi)H(P)}{|a_n|} \leq \Pi'|\xi - \alpha^{(\nu)}| \leq \frac{c_2(n,\xi)H(P)}{|a_n|},$$

where the prime indicates that the product is restricted to $|\xi - \alpha^{(\nu)}| \geq \psi$. (Hint: The $\xi - \alpha^{(\nu)}$ are roots of $P(\xi - x)$, so by Exercise 4.2.3,

$$\min(\psi^n, \psi^{-n}) 2^{-2n} L(P(\xi - x)) \le |a_n| \Pi' |\xi - \alpha^{(\nu)}| \le L(P(\xi - x)).$$

Now use Exercise 4.2.2.)

<u>4.2.5</u>. Show that, for $0 < \psi < \infty$ and $\xi \in \underset{\sim}{C}$ (e.g., by Exercise 4.2.4),

$$c_1(n,\xi) \frac{|P(\xi)|}{H(P)} \le \Pi'' |\xi - \alpha^{(\nu)}| \le c_4(n,\xi,\psi) \frac{|P(\xi)|}{H(P)},$$

where the double prime indicates that the product is restricted to $|\xi - \alpha^{(\nu)}| < \psi$.

<u>4.2.6</u>. Say P has no multiple roots. Show that

$$\frac{c(n)}{H^{n-1}} \le |\Pi' [\alpha^{(i)} - \alpha^{(j)}]| \le \frac{c(n) H^{n-1}}{|a_n|^{n-1}},$$

where Π' denotes a product over some subset of $S = \{(i,j) | 1 \le i < j \le n\}$. (Hint: The right-hand side is immediate from Theorem 4.1.2. Hence $H(D(x)) \le c(n) H^{n-1}$, where

$$D(x) = a_n^{n-1} \prod_{1 \le i < j \le n} \{x - [\alpha^{(i)} - \alpha^{(j)}]\}.$$

Clearly

$$(-x)^{n(n-1)/2} D(\tfrac{1}{x}) = a_n^{n-1} \prod_{1 \le i < j \le n} [\alpha^{(i)} - \alpha^{(j)}] \prod_{1 \le i < j \le n} [x - \frac{1}{\alpha^{(i)} - \alpha^{(j)}}].$$

Call the last factor on the right $E(x)$. By parts (a) and (b) of Exercise 4.2.1,

$$H(E(x)) \le H((-x)^{n(n-1)/2} D(\tfrac{1}{x})) = H(D(x)) \le c(n) H^{n-1},$$

so the left-hand side also follows from Theorem 4.1.2 (see Schneider [103], pp. 78-81.)

4.2.7. Say α is the root of P closest to $\xi \in \underset{\sim}{C}$. Then the left-hand side of

$$\frac{c(n,\xi)|P(\xi)|}{H(P)} \leq |\xi - \alpha| \leq \frac{c(n)H^{n-1}|P(\xi)|}{|a_n|}$$

is valid, and the right-hand side is valid if P has no multiple zeros. (Hint: The left-hand side follows easily from Exercise 4.2.5. For $i > 1$,

$$\frac{1}{2}|\alpha - \alpha^{(i)}| \leq |\xi - \alpha^{(i)}|,$$

so

$$2^{-n+1} \prod_{i>1} |\alpha - \alpha^{(i)}| \leq \prod_{i>1} |\xi - \alpha^{(i)}|.$$

Multiply by $|\xi - \alpha|$ and use Exercise 4.2.6.)

I do not know whether the right-hand side remains valid when H^{n-1} is replaced by H^{n-2}.

4.2.8. Show that $H(P(x)/(x - \alpha)) \leq nH(P)$. (Hint: Let $Q(x) = P(x)/(x - \alpha) = \beta_{n-1}x^{n-1} + \cdots + \beta_0$. Then

$$\beta_{n-1} = a_n, \quad \beta_{n-2} = a_{n-1} + \alpha a_n, \ldots, \quad \beta_0 = a_1 + \alpha a_2 + \cdots + \alpha^{n-1}a_n.$$

Thus if $|\alpha| \leq 1$, $\max|\beta_i| \leq nH$. If $|\alpha| > 1$, show that

$$H(Q(x)) = H\left(\frac{1}{\alpha}\frac{y^n P(1/y)}{y - (1/\alpha)}\right).)$$

4.2.9. Let $\beta \in \underset{\sim}{A}$, $d(\beta) = m$, and $P(\beta) \neq 0$. Show that

(a) $c(\beta)^n/H(P)^{m-1} < |P(\beta)|$, $\beta \in \underset{\sim}{R}$.

(b) $c(\beta)^n/H(P)^{(m-2)/2} < |P(\beta)|$, $\beta \in \underset{\sim}{C} - \underset{\sim}{R}$.

(Hint: For (a) the symmetric function theorem and (2.1.4) yield $1 \leq \Pi_{i=1}^{m} |b_m^n P(\beta^{(i)})|$, where b_m is the lead coefficient of β and the $\beta^{(i)}$ are the conjugates of β. Now

$$|P(\beta^{(i)})| \leq H\{1 + n[mH(\beta)]^n\} \leq c_1(\beta)^n H.)$$

4.2.10. Let $\beta \in \underset{\sim}{A}$, $d(\beta) = m$. Show that if $\alpha \neq \beta$, then

$$c(\beta)^n/H(\alpha)^m < |\beta - \alpha|.$$

(Hint: When β is not a conjugate of α, Exercise 4.2.8 yields

$$|a_n(\beta - \alpha^{(2)})\cdots(\beta - \alpha^{(n)})]$$

$$\leq nH\{1 + (n - 1)[mH(\beta)]^{n-1}\} \leq c(\beta)^n H,$$

and the result follows from Exercise 4.2.9.)

We now define

$$\omega_n(\xi, H) = \min|P(\xi)|$$

and

$$\omega_n^*(\xi, H) = \min|\xi - \alpha|,$$

where the first minimum is over all $P \in \underset{\sim}{Z}[x]$ such that $d(P) \leq n$, $H(P) \leq H$, $P(\xi) \neq 0$, and the second is over all $\alpha \in \underset{\sim}{A}$ such that $d(\alpha) \leq n$, $H(\alpha) \leq H$, and $\xi - \alpha \neq 0$. We also define $\omega_n(\xi)$ and $\omega_n^*(\xi)$ to be the largest real numbers such that for every $\varepsilon > 0$,

$$0 < |P(\xi)| \leq 1/H(P)^{\omega_n - \varepsilon},$$

and

$$|\xi - \alpha| \leq 1/H(\alpha)^{1+\omega_n^*-\varepsilon}$$

hold for infinitely many $P \in \underset{\sim}{Z}[x]$, $d(P) \leq n$, and infinitely many $\alpha \in \underset{\sim}{A}$, $d(\alpha) \leq n$, respectively.

A famous problem is to classify the real or complex numbers according to how well they can be approximated, that is, according to how fast $\omega_n(\xi,H)$ or $\omega_n^*(\xi,H)$ decreases. Thus one has the algebraic numbers and the Liouville numbers at opposite extremes. Mahler introduced a classification into A, S, T, and U numbers according to the rate of decrease of $\omega_n(\xi,H)$, and Koksma [50] later introduced an A^*, S^*, T^*, and U^* classification based on $\omega_n^*(\xi,H)$. That the two classifications are completely identical was finally shown by Wirsing [124]. Schmidt [59] showed that all these classes are nonempty by deducing the existence of T numbers from Theorem 6.1.2, and Sprindžuk [116] established Mahler's conjecture that the S numbers of "type 1" have measure 1. But if one is given a number [i.e., a Cauchy sequence from $\underset{\sim}{Q}(i)$], there seems to be little hope in general of determining to which class it belongs.

<u>4.2.11</u>. Show that

(a) $$\omega_n(\xi) = \limsup_H \frac{-\log \omega_n(\xi,H)}{\log H}.$$

(b) $$\omega_n^*(\xi) = \limsup_H \frac{-\log[H\omega_n^*(\xi,H)]}{\log H}.$$

(Hint: Show that lim sup $\geq \omega_n(\xi)$ is true and lim sup $> \omega_n(\xi)$ is false.)

4.2.12. Let $\beta \in A$, $d(\beta) = m$. Show that

(a) $\omega_n(\beta) \leq m - 1$, $\beta \in R$.

(b) $\omega_n(\beta) \leq \frac{m-2}{2}$, $\beta \in C - R$.

(c) $\omega_1(\xi_0) = \infty$ if $\xi_0 = \Sigma_{m=0}^{\infty} 10^{-m!}$.

(Hint: Use Exercises 4.2.9 and 4.2.11 for (a) and (b). Then show that

$$\omega_1(\xi_0, 2\cdot 10^{n!}) \leq \sum_{m=n}^{\infty} 10^{-n\cdot m!}.)$$

4.2.13. (a) If ξ is real, show that there is a nontrivial $P \in Z[x]$ with $d(P) \leq n$ and $H(P) \leq H$ such that $|P(\xi)| < c(n,\xi)/H^n$.

(b) If ξ is complex, show that the same holds with $|P(\xi)| < c(n,\xi)/H^{(n-1)/2}$. (Hint: In (a) use the pigeonhole principle (2.2.4) with $M = 1$, $\underline{x} = (a_0,\ldots,a_n)$, $N = n + 1$, $B = H$, and $L_i = 1 + |\xi| + \cdots + |\xi|^n$ to obtain

$$|P(\xi)| \leq \frac{2L_i H}{(H+1)^{n+1} - 2}.$$

In (b) use the pigeonhole principle in complex form (Exercise 2.2.3).)

4.2.14. If $d(\xi) \geq n$, show that

(a) $\omega_n(\xi) \geq n$, $\xi \in R$.

(b) $\omega_n(\xi) \geq (n-1)/2$, $\xi \in C$.

4.2.15. If $\xi \in C$, show that $c(n,\xi)\omega_n(\xi,H) \leq H\omega_n^*(\xi,H)$ and (hence) $\omega_n^*(\xi) \leq \omega_n(\xi)$. (Hint: Use the left-hand side of Exercise 4.2.7.)

4.2.16. For any $\xi \in \underset{\sim}{C}$ and $n, H \in \underset{\sim}{Z}^+$ show that

$$c(n)H^{-n+r} \prod_{i=1}^{r} \omega_{n_i}^*(\xi, H_i) \leq \omega_n(\xi, H),$$

where $1 \leq r \leq n$, $\Sigma_{i=1}^r n_i = n$, $c_1(n)H \leq \Pi_{i=1}^r H_i \leq c_2(n)H$, and all parameters except the c(n) are integral. (Hint: Choose $P(\xi)$ so that $\omega_n(\xi, H) = |P(\xi)|$. Say $P(x) = \Pi P_i(x)$, where the $P_i \in \underset{\sim}{Z}[x]$ are irreducible [they can be found by Gauss's lemma (Theorem 4.1.1)] and hence have no multiple zeros. Then from the right-hand side of Exercise 4.2.7 with $n_i = d(P_i)$ and $H_i = H(P_i)$, we have

$$\omega_{n_i}^*(\xi, H_i) \leq c(n_i)H_i^{n_i-1} |P_i(\xi)|.$$

(See Schneider [103], pp. 81 and 82).)

4.2.17. Show that

$$0 < |P(\xi)| \leq \frac{1}{H(P)^{\omega_n(\xi)-\epsilon}}$$

holds for infinitely many irreducible $P \in \underset{\sim}{Z}[x]$, $d(P) \leq n$, no matter how small $\epsilon > 0$ is. (Hint: If not, we can fix $\delta > 0$ and $c(n) > 0$ with $0 < c(n) < 1$ such that $|Q(\xi)| > c(n,\xi)/H(Q)^{\omega_n(\xi)-\delta}$ for all irreducible $Q \in \underset{\sim}{Z}[x]$ with $d(Q) \leq n$ and $Q(\xi) \neq 0$. Let P be a typical polynomial satisfying the inequality of the problem and write $P(x) = \Pi_{i=1}^r P_i(x)$, where the $P_i \in \underset{\sim}{Z}[x]$ are irreducible (Gauss's lemma). Then

$$0 < c(n,\xi)H(P)^{-\omega_n+\delta} < c(n,\xi) \prod_{i=1}^{r} [H(P_i)]^{-\omega_n+\delta}$$

$$< \prod_{i=1}^{r} |P_i(\xi)| = |P(\xi)| \leq H(P)^{-\omega_n+\epsilon} .$$

Choose $\epsilon < \delta$ and let $H(P) \to \infty$.)

4.2.18. Show that there are infinitely many irreducible polynomials of fixed degree n. For fixed degree n is there always an irreducible polynomial of height $H \geq 2$? What if $H = 1$? (Hint: Use Eisenstein's criterion.)

4.2.19. Show that $\omega_n^*(\xi) \geq \omega_n(\xi) - n + 1$. (Hint: Use Exercise 4.1.8 together with the hint to Exercise 4.2.16; this probably explains Wirsing's comment that the above result occurs in Schneider's book in hidden form. To obtain Wirsing's proof number the roots $\alpha^{(1)},\ldots,\alpha^{(n)}$ of any irreducible $P \in \underset{\sim}{Z}[x]$ so that

$$d_i = |\xi - \alpha^{(i)}| \leq d_j = |\xi - \alpha^{(j)}|$$

for $i \leq j$. From $1 \leq |D(P)|^{1/2}$ [see Section 4.3 for the notation D(P)] it follows that

$$1 \leq |a_n^{n-1} \prod_{1\leq i<j\leq n} 2\max(d_i,d_j)|$$

and hence

$$|\xi - \alpha| = d_1 \leq (\prod_{i=1}^{n} d_i)|a_n^{n-1}|2^{n(n-1)/2} \prod{}' d_i^{n-2}$$

$$\leq |P(\xi)||a_n^{n-2}| \frac{c_2(n,\xi)H(P)^{n-2}}{|a_n|^{n-2}}$$

by Exercise 4.2.4; the prime indicates that the product is over all $d_i \geq 1$. Now apply Exercises 4.2.17 and 4.2.11.)

<u>4.2.20</u>. Let ξ be fixed, $d(\xi) > n$, and let $S \subseteq \mathbb{Z}[x]$ be an infinite set of irreducible polynomials of degree at most n. Show that there is a set T of infinitely many pairs of polynomials, $T = \{(P_i,Q_i) \mid i \in \mathbb{Z}^+\}$, such that $i \neq j$ implies $P_i \neq P_j$ and $Q_i \neq Q_j$, and such that for $(P,Q) \in T$

(a) $P \in S$, $d(Q) \leq n$, $H(Q) \leq c(n)H(P)$.

(b) P,Q are relatively prime.

(c) $|Q(\xi)| < c(n,\xi)/H(P)^n$, $\xi \in \mathbb{R}$,

$|Q(\xi)| < c(n,\xi)/H(P)^{(n-1)/2}$, $\xi \in \mathbb{C} - \mathbb{R}$.

(Hint: Show that there is some c(n) such that $H(Q) \leq c(n)H(P)$ for $Q \in \mathbb{Z}[x]$ with $d(Q) \leq n$ implies $P \nmid Q$. Use Exercise 4.2.13 with Q in place of P and c(n)H(P) in place of H. Then $H(Q) \to \infty$ as $H(P) \to \infty$.)

<u>4.2.21</u>. If ξ is real and $n < d(\xi)$, show that

(a) $\frac{1}{2}(\omega_n(\xi) + 1) \leq \omega_n^*(\xi)$.

(b) $\frac{1}{2}(n + 1) \leq \omega_n^*(\xi)$.

Deduce that $\xi \in \mathbb{R}$ is transcendental if and only if $\omega_n(\xi) \to \infty$ if and only if $\omega_n^*(\xi) \to \infty$ (Wirsing [124]). (Hint: Since (a) and Exercise 4.2.14 imply (b), and the final statement then follows from Exercise 4.2.12, it suffices to show (a). Let the set S of Exercise 4.2.20 consist of the irreducible polynomials found in Exercise 4.2.17. For $(P,Q) \in T$ (see Exercise 4.2.20) write

$$P(x) = a_n \prod_{i=1}^{n'} (x - \alpha^{(i)}), \quad Q(x) = b_m \prod_{j=1}^{m} (x - \beta^{(j)}),$$

and

$$R(P,Q) = a_n^m b_m^{n'} \prod_{i,j} (\alpha^{(i)} - \beta^{(j)}) \qquad (n',m \leq n),$$

where the roots of P,Q are ordered as in Exercise 4.2.19. Set $d_i = |\xi - \alpha^{(i)}|$ and $D_j = |\xi - \beta^{(j)}|$.

1. Say (i) $n' = 1$ a.a. or (ii) $m = 1$ a.a. Here a.a. (read "almost always") means "with only a finite number of exceptions." Then

$$c(\xi)H \leq |a_i| \leq H \quad \text{a.a.,}$$

where $H = H(P)[H(Q)]$ and a_i denotes either coefficient of $P(Q)$ in case (i) [case (ii)].

Case 1(i): Here $\omega_n \leq \omega_n^*$, so the result holds since $1 \leq \omega_n(\xi)$ by Exercise 4.2.14.

Case 1(ii): Here $n \leq \omega_n^*$, so if $\omega_n \leq 2n - 1$, the result is immediate. Otherwise $(\omega_n + 1)/2 \leq \omega_n - n + 1 \leq \omega_n^*$ by Exercise 4.2.19.

2. Say (i) $d_2 \geq 1$ a.a. or (ii) $D_2 \geq 1$ a.a.

Case 2(i): By Exercise 4.2.5 with $\psi = 1/2$,

$$|\xi - \alpha^{(1)}| \leq c(n,\xi) \frac{|P(\xi)|}{H(P)} \leq \frac{c(n,\xi)}{H(P)^{1+\omega_n-\varepsilon}},$$

and the result follows as in case 1(i).

Case 2(ii): By Exercises 4.2.4 and 4.20(a)

$$|\xi - \beta^{(1)}| \le \frac{c(n,\xi)}{H(Q)^{n+1}},$$

and since $H(Q) = H(\beta^{(1)})$, the result follows as in case 1(ii).

3. Say $n', m \ge 2$ and $d_2, D_2 < 1$ on an infinite subset of T. By reasoning similar to that in the hint to Exercise 4.2.19 [with R(P,Q) in place of D(P)] one obtains

$$(*) \qquad 1 \le |R(P,Q)| \le c(n,\xi)H(P)^m H(Q)^{n'} \prod_{d_i<1} d_i^{e_i} \prod_{D_j<1} D_j^{f_j},$$

where $e_i = |\{j \mid D_j < d_i\}| \le m$ and $f_j = |\{i \mid d_i \le D_j\}| \le n$. Clearly $e_1 \le e_2 \le \cdots$, $f_1 \le f_2 \le \cdots$, and if e_s, f_s both exist, exactly one of $e_s \ge s$, $f_s \ge s$ occurs. Now $e_1 \le 1$ is true except for finitely many $(P,Q) \in T$ since it follows from $e_1 \ge 2$, (*), and Exercises 4.2.5 and 4.2.20(a) that

$$1 \le c(n,\xi)H(P)^{2n-2}/H(P)^{2\omega_n - 2\varepsilon},$$

while $n \le \omega_n$ by Exercise 4.2.14. Similarly $f_1 \le 1$, except for finitely many $(P,Q) \in T$. Now show that the set equality

$$\{e_1, f_1, e_2, f_2\} = \{0,1,1,t\}$$

holds a.a. where $t \ge 2$. Then proceed as in Exercise 4.2.19 for each of the four possible distinct cases. For example, if the case $e_1 = 1$, $f_1 = 0$, $e_2 \ge 2$, and $f_2 = 1$ occurs infinitely often, then it follows from (*) and Exercise 4.2.5 that

$$D_1^2 \le d_1 D_1 \le c(n,\xi)H(P)^{m-2}H(Q)^{n'-1}|P(\xi)|^2|Q(\xi)|.$$

By Exercise 4.2.20,

$$|\xi - \beta^{(1)}|^2 \le c(n,\xi)/H(Q)^{2+\omega_n-n+(\omega_n+1)-2\epsilon}$$

so from $H(Q) = H(\beta^{(1)})$ and Exercise 4.2.14(a) it follows that $\omega_n^* \ge \frac{1}{2}(\omega_n + 1)$.)

<u>4.2.22</u>. If ξ is real and $n < d(\xi)$, show that

$$\omega_n^*(\xi) \ge \frac{\omega_n(\xi)}{\omega_n(\xi) - n + 1} .$$

(Hint: Let $\xi_i = \xi + i$, $0 \le i \le n$, and for each $B > 0$ and $\sigma > \omega_n(\xi) \ge n > 0$ find a nontrivial $P = a_n x^n + \cdots + a_0 \in \underset{\sim}{Z}[x]$ such that

$$|P(\xi)| < c(n,\xi)B^{-\sigma},$$

$$|P(\xi_i)| < c(n,\xi)B \qquad (1 \le i \le n - 1),$$

$$|P(\xi_n)| < c(n,\xi)B^{\sigma-n+1}.$$

[Use Exercise 2.2.4 with $(a_0,\ldots,a_n)$ for $\underline{x}$, $n + 1$ for N, $1 + |\xi_i| + \cdots + |\xi_i|^n$ for L_i, etc.] Note that as $B \to \infty$, infinitely many such P are obtained. Moreover, when B is large enough and $\sigma > \omega_n + 2\epsilon$, we have $H(P) \ge c(n,\xi)B^{1+\delta}$, $\delta = \delta(n,\xi) > 0$, since

$$H(P)^{-\omega_n-\epsilon} \le |P(\xi)| \le c(n,\xi)B^{-\sigma}.$$

Hence by Exercise 4.2.5, for $0 \le i \le n - 1$,

$$\prod_{\alpha} |\xi_i - \alpha| < c(n,\xi)B^{-\delta}.$$

Thus at least one zero of P is tending to ξ_i as $B \to \infty$ ($0 \leq i \leq n - 1$) and therefore exactly one. So for B sufficiently large, Exercise 4.2.5 yields

$$|\xi - \alpha| \leq c(n,\xi)/H(P)B^{\sigma};$$

we thus need a lower bound for B. Solve for the a_i in terms of the $P(\xi_i)$ by Cramer's rule (the determinant is Vandermondian) and obtain

$$H(P) \leq \max_i |a_i| \leq c(n,\xi)B^{\sigma-n+1}.)$$

The functions $\omega_n(\xi)$ and $\omega_n^*(\xi)$ are used to classify the real numbers $\underset{\sim}{R}$ into distinct types. Let $\omega = \limsup (\omega_n(\xi)/n)$ and let μ be the smallest integer n such that $\omega_n(\xi) = \infty$ ($\mu = \infty$ if no such integer exists). We say that ξ is an

A number if $\omega = 0$, $\mu = \infty$

S number if $0 < \omega < \infty$, $\mu = \infty$

T number if $\omega = \infty$, $\mu = \infty$

U number if $\omega = \infty$, $\mu < \infty$.

(The A numbers are algebraic, the S honors Carl Ludwig Siegel, and T, U follow in alphabetical order.) The use of $\omega_n^*(\xi)$ instead of $\omega_n(\xi)$ leads to the A^*, S^*, T^*, and U^* classification introduced by Koksma [50].

Each of these classes can be further subdivided in a natural way; for example, an S number ξ can be further classified by its type $\theta = \sup(\omega_n(\xi)/n)$.

4.2.23. Show that $A = A^*$, $S = S^*$, $T = T^*$, and $U = U^*$. (Hint: Show that A and A^* are precisely the algebraic numbers, then obtain the rest from Exercises 4.2.15 and 4.2.21.)

4.2.24. If $\beta \in \underset{\sim}{R}$ and $d(\beta) = m$, show that $\omega_{m-1}(\beta) = \omega^*_{m-1}(\beta) = m - 1$. (Hint: Use Exercises 4.2.12, 4.2.14, 4.2.15, and 4.2.22.)

4.2.25. If $\xi \in \underset{\sim}{R}$ and $n < d(\xi)$, show that

$$\omega^*_n(\xi) \geq \frac{(n + 2) + \sqrt{(n^2 + 4n - 4)}}{4}.$$

(Hint: Use Exercises 4.2.21 and 4.2.22.)

*4.2.26. If $\xi \in \underset{\sim}{C} - \underset{\sim}{R}$ and $n < d(\xi) = s \leq \infty$, show that

(a) $\omega^*_n(\xi) \geq \omega_n(\xi) - \frac{n - 1}{2}$.

(b) $\omega^*_n(\xi) \geq \frac{1}{2}\,\omega_n(\xi)$.

(c) $\omega^*_n(\xi) \geq \frac{\omega_n(\xi)}{2\omega_n(\xi) - n + 2}$.

(d) $\omega_{s-1}(\xi) = \omega^*_{s-1}(\xi) = \frac{s - 2}{2}$.

(e) $\omega^*_n(\xi) \geq \frac{n}{4}$.

(Hint: The same techniques apply to (a), (b) and (c); (d) and (e) are then easy consequences. See Wirsing [124].)

**4.2.27. Is $n \leq \omega^*_n(\xi)$?

4.2.28. If ξ is real and $n < d(\xi)$, show that there is an $\alpha \in \underset{\sim}{A}$ such that

$$|\xi - \alpha| \leq 1/H(\alpha)^{(n+3)/2 - \epsilon}.$$

(Hint: Use Exercise 4.2.21(b).)

4.2.29. If $\beta \in \underset{\sim}{R}$ and $d(\beta) = m$, show that

(a) $d(\alpha) < m$ implies that $|\beta - \alpha| \geq c_1(\beta)/H(\alpha)^m$.

(b) $|\beta - \alpha| \leq c_2(\beta)/H(\alpha)^m$ has infinitely many solutions α with $d(\alpha) < m$.

(Hint: (a) follows from Exercise 4.2.10; (b) almost follows from Exercise 4.2.24, but "misses by an ε".)

*4.2.30. Show that almost all numbers are S numbers of type $\theta \leq 1$. (Hint: Show first that the S^* numbers of type $\theta^* > 1$ have measure zero (use the well-known Borel-Cantelli lemma). The same argument applies to the T^* and U^* numbers.)

4.3. Discriminants and Resultants

This section gives estimates for discriminants and resultants. It may be omitted on a first reading.

Definition 4.3.1. Let $P, Q \in \underset{\sim}{C}[x]$ where

$$P(x) = a_n x^n + \cdots + a_0 = a_n(x - \alpha_1)\cdots(x - \alpha_n)$$

and

$$Q(x) = b_m x^m + \cdots + b_0 = b_m(x - \beta_1)\cdots(x - \beta_m).$$

We define $D(P)$, the discriminant of P, by

$$D(P) = a_n^{2n-2} \prod_{i<j} (\alpha_i - \alpha_j)^2. \tag{4.3.1}$$

The resultant $R(P,Q)$ of P and Q is defined by

$$R(P,Q) = a_n^m b_m^n \prod_{i,j} (\alpha_i - \beta_j). \tag{4.3.2}$$

If $P,Q \in \underset{\sim}{Z}[x]$, then by Lemma 4.1.1, $D(P)$ and $R(P,Q)$ are algebraic integers; hence by the symmetric function theorem they are rational integers. Clearly $P'(\alpha_i) \neq 0$, $1 \leq i \leq n$, if $D(P) \neq 0$ and it is also true that $P(\beta_j) \neq 0$ for $1 \leq j \leq m$ if $R(P,Q) \neq 0$. These last results can be stated quantitatively.

Theorem 4.3.1. If $D(P) \neq 0$, then

$$|D(P)| \leq 2^{n(n-1)} L_2^{2n-2}(P) \qquad (n \geq 2) \tag{4.3.3}$$

and

$$2^{-\binom{n-1}{2}} |D(P)|^{1/2} L_2(P)^{-n+2} \max(1, |\alpha_i|)^{n-2} < |P'(\alpha_i)| \qquad (n \geq 2). \tag{4.3.4}$$

If $R(P,Q) \neq 0$, then

$$|R(P,Q)| \leq L^m(P) L_2^n(Q) \tag{4.3.5}$$

and

$$|R(P,Q)| L(P)^{-m+1} L_2(Q)^{-n} \max(1, |\beta_j|)^n \leq |P(\beta_j)|. \tag{4.3.6}$$

Proof. Inequality (4.3.3) follows from Lemma 4.2.2 and also (4.3.4):

$$|D(P)| = |a_n^{2n-2} \prod_{i<j} (\alpha_i - \alpha_j)^2|$$

$$\leq |a_n^{2n-4}| |P'(\alpha_k)|^2 \, 2^{2\binom{n}{2}-2(n-1)} \prod_{i\neq k} \max(1, |\alpha_i|)^{2n-4}$$

$$\leq \frac{2^{2\binom{n-1}{2}} |P'(\alpha_k)|^2 L_2(P)^{2n-4}}{\max(1, |\alpha_k|)^{2n-4}} .$$

Inequalities (4.3.5) and (4.3.6) are obtained similarly:

$$|R(P,Q)| = |b_m^n \prod_{j=1}^{m} P(\beta_j)| \le L^m(P)|b_m^n \prod_{j=1}^{m} \max(1,|\beta_j|)^n|$$

$$\le L^m(P)L_2^n(Q)$$

and

$$|R(P,Q)| \le |P(\beta_k)||b_m^n|L^{m-1}(P) \prod_{j\ne k} \max(1,|\beta_j|)^n$$

$$\le \frac{|P(\beta_k)|L^{m-1}(P)L_2^n(Q)}{\max(1,|\beta_k|)^n} .$$

□

In (4.3.3) and (4.3.4) the constants $2^{n(n-1)}$ and $2^{-\binom{n-1}{2}}$ can be improved to n^n and $(n-1)^{-(n-1)/2}$, respectively (see Exercise 4.3.16). For the results of this section see Kasch and Volkman [49], p. 444. For further results on discriminants and resultants see Güting [43].

Exercises

Problems 4.3.1 through 4.3.4 are modifications of lemmas that may be found in Sprindžuk [117], pp. 19-25. Let $P \in \mathbb{C}[x]$, $d(P) = n \ge 2$, and $P = a_n x^n + \cdots + a_0 = a_n(x - \alpha_1)\cdots(x - \alpha_n)$. For $\xi \in \mathbb{C}$ choose α so that $P(\alpha) = 0$ and $|\xi - \alpha|$ is minimal.

<u>4.3.1</u>. Show that

$$|\xi - \alpha|^r \le 2^{n-r}|\alpha - \alpha_1|\cdots|\alpha - \alpha_{r-1}| \frac{|P(\xi)|}{|P'(\alpha)|}$$

for $r = 1,\ldots,n$ if $\alpha \ne \alpha_i$, $i = 1,\ldots,r - 1$. (Hint:

$$|P'(\alpha)| = |a_n|\Pi'|\alpha - \alpha_j| \leq 2^{n-r}|a_n||\alpha - \alpha_1|\cdots|\alpha - \alpha_{r-1}|\Pi''|\xi - \alpha_j|$$

and

$$|a_n|\Pi''|\xi - \alpha_j| = \frac{|P(\xi)|}{|\xi - \alpha||\xi - \alpha_1|\cdots|\xi - \alpha_{r-1}|}.)$$

4.3.2. Show that

$$|\xi - \alpha| \leq 2^{\binom{n}{2}} L_2^{n-2}(P)|D(P)|^{-1/2}|P(\xi)|\max(1, |\alpha|)^{-n+2}.$$

(Hint: Use Exercise 4.3.1 and Theorem 4.3.1.)

4.3.3. If $|P(\xi)| \leq L_2(P)^{-\omega}$, show that $(n \geq 3)$

$$|\xi - \alpha| \leq 2^{\frac{1}{3}\binom{n}{2}} L_2^{-\frac{1}{3}(2\omega-n)-1}(P)|D(P)|^{-1/6}\max(1, |\alpha_j|)^{1-\frac{1}{3}n},$$

where $|\xi - \alpha_j|$ is minimal among the $|\xi - \alpha_k|$ for which $\alpha_k \neq \alpha$. (Hint: Let $P(x) = (x - \alpha)P_\alpha(x)$, so $|D(P)| = |P'(\alpha)|^2|D(P_\alpha)|$. Then by Exercise 4.3.1 with $r = 1$,

$$|\xi - \alpha| \leq 2^{n-1}L_2^{-\omega}(P)|D(P)|^{-1/2}|D(P_\alpha)|^{1/2},$$

and by Exercise 4.3.2 applied to $P_\alpha(x)$,

$$|\xi - \alpha_j| \leq 2^{\binom{n-1}{2}} L_2^{n-3}(P)|P_\alpha(\xi)|\max(1, |\alpha_j|)^{-n+3}|D(P_\alpha)|^{-1/2}.$$

Multiply both sides by $|\xi - \alpha|$.)

4.3.4. (A partial converse to Exercise 4.3.1.) Assume that $|P(\xi)| \leq L_2^{-\omega}(P)$ for $\omega \geq n - 1 \geq 2$ and

$$c_1(n,\lambda)\,|D(P)|^{-1/2}\max(1,|\alpha_j|)^{(3-n)/2} \le L_2(P),$$

where $\lambda > 0$ and α_j is chosen as in Exercise 4.3.3. Let $|\alpha_q - \alpha|$ be minimal among the $|\alpha_k - \alpha_q|$, $\alpha_k \ne \alpha$. Show that

(a) $c_2(n,\lambda)\,|P(\xi)|/|P'(\alpha)| \le |\xi - \alpha|$ when $|\xi - \alpha| \le \lambda|\alpha - \alpha_q|$.

(b) $(c_3(n,\lambda)\,|\alpha - \alpha_q|\,|P(\xi)|/|P'(\alpha)|)^{1/2} \le |\xi - \alpha|$ otherwise.

Here $c_1(n,\lambda) = (1 + \lambda)^{1/2}\lambda^{-1}2^{\binom{n}{2}-\frac{3}{2}}$, $c_2(n,\lambda) = (1 + \lambda)^{-n+1}$, and $c_3(n,\lambda) = \lambda(1 + \lambda)^{-1}2^{-n+2}$. [If $P \in \mathbb{Z}[x]$ and P has no multiple zeros, then $1 \le |D(P)|$, so the postulated lower bound for $L_2(P)$ can be replaced by (a possibly larger) one that depends _only_ on n and λ.] (Hint: For (a), $|\xi - \alpha_k| \le |\xi - \alpha| + |\alpha - \alpha_k| \le (\lambda + 1)|\alpha - \alpha_k|$, so

$$|P'(\alpha)| = |a_n|\Pi'|\alpha - \alpha_k| \ge |a_n|(1 + \lambda)^{-n+1}\Pi'|\xi - \alpha_k|$$

$$= (1 + \lambda)^{-n+1}\frac{|P(\xi)|}{|\xi - \alpha|}.$$

For (b), let $\alpha_k \ne \alpha,\alpha_q$. If $|\alpha_k - \alpha| < |\xi - \alpha|$, it would follow by Exercise 4.3.3 that

$$|\alpha - \alpha_q||\alpha - \alpha_k||\alpha_q - \alpha_k| \le \lambda^{-1}|\xi - \alpha|^2(1 + \lambda^{-1})|\xi - \alpha|$$

$$\le (1 + \lambda)\lambda^{-2}2^{\binom{n}{2}}L_2^{-n-1}(P)\,|D(P)|^{-1/2}\max(1,|\alpha_j|)^{3-n},$$

and since $|D(P)|^{1/2} \le |\alpha - \alpha_q||\alpha - \alpha_k||\alpha_q - \alpha_k|2^{\binom{n}{2}-3}L_2^{n-1}(P)$, this contradicts the lower bound for $L_2(P)$. Thus

$$|\xi - \alpha_k| \le |\xi - \alpha| + |\alpha_k - \alpha| \le 2|\alpha_k - \alpha|,$$

so

$$|P'(\alpha)| = |a_n|\Pi'|\alpha_k - \alpha| > 2^{-n+2}|\alpha - \alpha_q|\Pi'|\xi - \alpha_k|$$

$$= \frac{2^{-n+2}|\alpha - \alpha_q||P(\xi)|}{|\xi - \alpha||\xi - \alpha_q|}.$$

Since

$$|\xi - \alpha_q| \le |\xi - \alpha| + |\alpha - \alpha_q|$$

$$\le (1 + \lambda^{-1})|\xi - \alpha|,$$

the result follows.)

Exercises 4.3.5 through 4.3.19 are related to M. Fekete's concept of the "transfinite diameter." Exercises 4.3.6 through 4.3.9 are from Cassels [19], pp. 1-8. Throughout we let $\alpha_1,\ldots,\alpha_n$ denote all conjugates of the algebraic number α of degree n.

<u>4.3.5</u>. In terms of the coefficients of its minimal polynomial, how far apart can two algebraic conjugates of α be? How large can $|(\alpha - \alpha_1)\cdots(\alpha - \alpha_k)|$ be, $1 \le k \le n$? (Hint: We have $|(\alpha - \alpha_1)\cdots(\alpha - \alpha_k)| \le 2^k\Pi_1^k$, where Π_1 denotes the product over conjugates of α of modulus at least 1.)

<u>4.3.6</u>. Given $0 < x_j \le 1 + \varepsilon$, where $0 < \varepsilon$ and $\Pi_{j=1}^n x_j = 1$, show that

$$\prod_{j=1}^{n} |x_j - 1| \le (C\varepsilon)^n$$

for some absolute constant C. (Hint: Show that if $x_1 < x_2 < 1$ or $1 < x_1 < x_2$, then $0 < (x_1 - 1)(x_2 - 1) \le (\sqrt{x_1x_2} - 1)^2$, whereas if $x_1 < 1 < x_2 < t$, then $(x_2 - 1)(1 - x_1) < (t - 1)(1 - x_1x_2/t)$.)

4.3.7. If $|z_i| = 1$, $i = 1,\ldots,n$, show that

$$D^2 = \left|\prod_{i<j} (z_i - z_j)\right|^2 \le n^n.$$

(Hint: D is the absolute value of a Vandermondian determinant. Estimate it by Hadamard's inequality.)

4.3.8. Say $1 \le \rho$ and $z_1,\ldots,z_n$ are complex numbers such that $|z_i| \le \rho$. Show that

$$\prod_{i\ne j} |z_i\bar{z}_j - 1| \le \rho^{2n(n-1)}n^n.$$

(This modified form of an inequality of Cassels, as well as the hint below, are due to J. Ralph Alexander.) (Hint: Consider

$$\Pi = \prod_{i\ne j} |z_i\bar{z}_j - \omega_i\bar{\omega}_j| \qquad (i,j = 1,\ldots,n),$$

where $|z_i|, |\omega_i| \le \rho$. Here Π is the absolute value of a polynomial in z_i (ω_i), so by the maximum-modulus principle it will be maximal when $|z_i| = \rho (|\omega_i| = \rho)$. Moreover,

$$\Pi = \prod_{i\ne j} |\omega_i\omega_j| \prod_{i\ne j} |u_i\bar{u}_j - 1|,$$

where $|u_i| = |z_i/\omega_i| = 1$. Apply Exercise 4.3.7.)

4.3.9. Show that there is an absolute constant $c > 0$ such that if $|\alpha_j| < 1 + c/n$, $1 \le j \le n$, where α is an algebraic integer, then $\{\alpha_1,\ldots,\alpha_n\} = \{\alpha_1^{-1},\ldots,\alpha_n^{-1}\}$. (Hint: Choose c so that $\Pi|\alpha_j| < 2$. Then $m \in \underset{\sim}{Z}$, where

$$m = \prod_{j,k} |\alpha_j\bar{\alpha}_k - 1| = \prod_j |\alpha_j\bar{\alpha}_j - 1| \prod_{j\neq k} |\alpha_j\bar{\alpha}_k - 1|$$

$$\leq \left(\frac{c_1}{n}\right)^n n^n \left(1 + \frac{c_2}{n}\right)^{2n(n-1)} \leq (c_1 e^{c_3})^n$$

by Exercises 4.3.6 and 4.3.8. Make $c_1 e^{c_3} < 1$.)

**<u>4.3.10</u>. Given $1 \leq r < R$, show that there are infinitely many algebraic integers α such that $r \leq |\alpha_j| \leq R$, $1 \leq j \leq n$. (Hint: See Fekete and Szegö [32], p. 171, Theorem K.)

<u>4.3.11</u>. Given any n points in the complex plane, show that there is a polynomial $P \in \underset{\sim}{Z}[x]$ of degree n whose roots are arbitrarily close to these points. (*How small can L(P) remain if each root is to be approximated within ϵ?)

**<u>4.3.12</u>. (a) Is there an absolute constant $c > 0$ such that if the algebraic integer α satisfies $|\alpha_j| < 1 + c/n$, $1 \leq j \leq n$, then α is a root of unity? (This question was proposed by Schinzel and Zassenhaus [97], who obtained the conclusion under the stronger hypothesis that $|\alpha_j| < 1 + 2^{-2s-4}$, $1 \leq j \leq n$, where 2s is the number of complex conjugates of α.)

(b) Show that there is an absolute constant $c > 0$ such that if

$$|\alpha_j| \leq 1 + c(n^2 \ln 6n)^{-1}$$

for $1 \leq j \leq n$, then α is a root of unity. (This was proved by Blanksby and Montgomery [12].)

For the following exercises let U denote the unit circle and $I = [-2,2]$.

4.3.13. Show that if all the conjugates of an algebraic integer lie on U, then α is a root of unity, whereas if they all lie in I, then $\alpha = 2 \cos 2\pi p/q$, where $p,q \in \mathbb{Z}$ (Kronecker). (Hint: For the first part show that the coefficients of $(x - \alpha_1^m)\cdots(x - \alpha_n^m)$ are bounded independently of $m \in \mathbb{Z}$. For the second part note that if $f(z) = z + z^{-1}$, then $f(U) = I$.)

If S is a compact set in the plane, we define $d_n = d_n(s)$, $n \geq 2$, by

$$d_n^{\binom{n}{2}} = \max \prod_{j<k} |z_j - z_k| ,$$

where $z_1,\ldots,z_n$ vary independently over S. Clearly d_2 is just the diameter of S.

4.3.14. Show that $d_n \geq d_{n+1} \geq 0$, $n \geq 2$. (Hint: Find $z_1,\ldots,z_{n+1} \in S$ such that

$$d_{n+1}^{\binom{n+1}{2}} = \prod_{j<k} |z_j - z_k|.$$

Then, for fixed k,

$$d_{n+1}^{\binom{n+1}{2}} \leq d_n^{\binom{n}{2}} \prod_{j\neq k} |z_j - z_k|.$$

Multiply all these inequalities together.)

Thus $d_\infty = d_\infty(S) = \lim_{n\to\infty} d_n(S)$ exists. We call d_∞ the transfinite diameter of S.

4.3.15. Fix an integer m. Show that if S contains all roots of infinitely many distinct irreducible polynomials $P(x) \in \underset{\sim}{Z}[x]$ with lead coefficient m, then $d_\infty(S) \geq 1$. (Hint: Consider the discriminants of the P(x). This result shows that $d_\infty(U) \geq 1$ and $d_\infty(I) \geq 1$.)

4.3.16. Show that (a) $d_\infty(U) = 1$ and (b) $d_\infty(I) = 1$. Show that (4.3.3) of Theorem 4.3.1 can be replaced by

$$(4.3.3)^* \qquad |D(P)| \leq n^n L_2^{2n-2}(P).$$

(Hint: For (a) use Exercise 4.3.8; (b) is harder; for a direct proof see Schur [104], pp. 377-380. Further study of transfinite diameters requires the use of Chebyshev polynomials. For a proof of (b) based on their properties see Hille [48], pp. 264-270.)

4.3.17. (a) Let $\zeta, \zeta_1, \ldots, \zeta_n$ be points on the unit circle and set

$$f(\zeta) = \prod_{j=1}^{n} |\zeta - \zeta_j|^2.$$

Show that the average value of $f(\zeta)$ is least when the ζ_i are uniformly spaced.

(b) If $\varphi_1, \ldots, \varphi_n$ are real, show that

$$\int_0^\pi \cos^2(\varphi - \varphi_1) \cdots \cos^2(\varphi - \varphi_n)\, d\varphi \geq \frac{\pi}{2^{2n-1}} .$$

*4.3.18. Show that if a line segment has length greater than 4, then it contains all roots of infinitely many distinct monic irreducible polynomials $P(x) \in \underset{\sim}{Z}[x]$. [It is not known whether this holds for line segments of length exactly 4 with nonintegral end points.] (Hint: See R. M. Robinson [91], pp. 305-315; his proof relies on the properties of Chebyshev polynomials.)

4.3.19. Let $h(x + \pi) = h(x)$, $h(-x) = h(x)$, and $h[(\pi/2) - x] = h[(\pi/2) + x]$, and suppose that $g[h(x)]$ is a strictly convex (concave) function of x for $0 \leq x \leq \pi$. Let

$$f(\theta_1,\theta_2) = f(e^{i\theta_1},e^{i\theta_2}) = g[h((\theta_1-\theta_2)/2)].$$

Show that $\sum_{1\leq i<j\leq n} f(\theta_i,\theta_j)$ is minimal (maximal) if and only if the $e^{i\theta_j}$, $1 \leq j \leq n$, are the vertices of a regular n-gon. (Hint: A proof can be based on the methods of Fejes Tóth [31]. This yields a slightly more precise result than Exercise 4.3.16(a) on taking $h(x) = |\sin x|$, $g(x) = \log x$, and using the well-known fact (see Pollard [84], pp. 56 and 57; replace p by n) that the product of the squares of the lengths of all diagonals of the regular n-gon inscribed in the unit circle is n^n.)

*4.3.20. Given nonnegative integers s,t, is there an irreducible $P(x) \in \mathbb{Z}[x]$, $d(P) = n = s + 2t$, that has s real roots and t pairs of conjugate imaginary roots?

4.3.21. Let $K = \mathbb{Q}(\alpha)$, where the conjugates of α are $\alpha_1,\ldots,\alpha_n$. Show that the integers of K are (a) not dense anywhere if $d(\alpha) = 1$ or $d(\alpha) = 2$ and α_1,α_2 are complex, (b) dense in $\mathbb{R}$ if $d(\alpha) \geq 2$ and $\alpha_1,\ldots,\alpha_n$ are real, and (c) dense in $\mathbb{C}$ if $d(\alpha) \geq 3$ and α_i is complex for some i, $1 \leq i \leq n$.

*4.3.22. Let the algebraic integer α have conjugates $\alpha = \alpha_1,\alpha_2,\ldots,\alpha_n$. Let β be algebraic and

$$\Lambda = \{\lambda \,|\, \Pi' |\alpha_i| = \lambda L_2(\alpha), \text{ where } \alpha \in \underset{\sim}{Q}(\beta)\},$$

where Π' denotes the product over $|\alpha_i| \geq 1$ (see Lemma 4.2.2). What can be said about the set of limit points of Λ?

4.3.23. Let $P(x) = a_n x^n + \cdots + a_0 = a_n(x - \alpha_1)\cdots(x - \alpha_n)$ be the minimal polynomial of α. Define den α, the denominator of α, to be the smallest positive integer d such that $d\alpha$ is an algebraic integer. Show that den $\alpha | a_n$, but that den $\alpha \neq a_n$ in general. Define

$$\text{size}(\alpha) = \max(\log \text{den } \alpha, \log|\alpha_1|, \ldots, \log|\alpha_n|).$$

Show that $-2n \text{ size}(\alpha) \leq \log|\alpha_i|$ for $1 \leq i \leq n$. (Hint: Consider $P(x) = 3^4x^2 + 3^2x + 1$. The inequality for size(α) is stated in Lang [55], p. 642.)

4.3.24. (Spira [115], p. 192, Problem E 2217). If $P, Q \in \underset{\sim}{C}[x]$ and $d(P) = d(Q) = 1$, show that

$$\tfrac{1}{2}(-1 + \sqrt{5})H(P)H(Q) \leq H(PQ).$$

When does equality hold?

4.3.25. Let $P(x) = x^n + a_0$ and $Q(x) = x^2 + b_1x + b_0$, where $a_0, b_0, b_1 \in \underset{\sim}{R}$. Show that the following inequalities are best possible:

$$(2 + \sqrt{2})^{-1}L_2^2(P)L_2^2(Q) \leq L_2^2(PQ) \qquad (n = 1).$$

$$L_2^2(P)L_2^2(Q) \leq 2L_2^2(PQ) \qquad (n = 2).$$

$$= L_2^2(PQ) \qquad (n = 3).$$

4.3.26. If $d(P_1) = d(P_2) = 1$, show that

$$2^{-1/\mu}L_\mu(P_1)L_\mu(P_2) \le L_\mu(P_1P_2) \qquad (0 \le \mu \le 2)$$

for $P_1, P_2 \in \underset{\sim}{C}[x]$ is true and best possible.

4.3.27. If $P(x) = a_n(x - \alpha^{(1)})\cdots(x - \alpha^{(n)})$, show that

$$H\left(\frac{P(x)}{(x - \alpha^{(1)})\cdots(x - \alpha^{(m)})}\right) \le \min\left[n^m, (n + 1)\binom{n-m}{\left[\frac{n-m}{2}\right]}H(P)\right].$$

4.3.28. Let $f_\mu(m,n) = \sup|a_n\alpha^{(1)}\cdots\alpha^{(m)}/L_\mu(P)|$, where

$$P(x) = a_nx^n + \cdots + a_0 = a_n(x - \alpha^{(1)})\cdots(x - \alpha^{(n)})$$

and $1 \le m \le n$. Show that

(a) $f_\infty(1,1) = f_\infty(1,2) = f_\infty(n,n) = 1$.

(b) $f_\infty(1,n) \le 2$.

(c) $f_\mu(m,n) \le 1$ if $\mu \le 2$.

(Hint: In (b), $|a_n\alpha| \le |a_n|(1 + H/|a_n|) \le 2H$ by $(3.2.1)^*$.)

4.3.29. When does inequality hold in Lemma 4.2.2?

4.3.30. Show that there is some $\lambda > 0$ (e,g., $\lambda = 1/8$) such that

$$1 + |x + y + z| + |xy + xz + yz| + |xyz| \ge \lambda(|x| + |y| + |z|).$$

4.3.31. Obtain Ostrowski's generalization of Lemma 4.2.2, namely, that for $\mu \ge 2$ we have

$$|\cdots|^{\mu} + \cdots + |\cdots|^{\mu} \le \left(\frac{L_2(P)}{|a_n|}\right)^{\mu} + k - 2,$$

where each of the k terms $|\cdots|$ on the left denotes a nonempty product of distinct roots of P that do not occur in any other $|\cdots|$. The case $\mu < 2$ seems to be open.

Let $\alpha \in \underset{\sim}{A}$. Define the height of a field $\underset{\sim}{K} = \underset{\sim}{Q}(\alpha)$ by $H(\underset{\sim}{K}) = \min\{H(\beta) \mid \underset{\sim}{Q}(\beta) = \underset{\sim}{K}\}$. For example, $H(Q(\sqrt{5})) = 1$ since $\underset{\sim}{Q}(\sqrt{5}) = \underset{\sim}{Q}(\beta)$, where β is the root of $x^2 + x - 1 = 0$.

*<u>4.3.32</u>. Does $H(\underset{\sim}{Q}(\alpha))$ depend effectively on α?

*<u>4.3.33</u>. For $\alpha \in \underset{\sim}{A}$, does $H(\underset{\sim}{Q}(\alpha))$ assume arbitrarily large values? Does it assume all integral values?

Chapter 5

THUE'S THEOREM

> Nach der vorangegangenen Theorie der Pellschen Gleichung wird es dem Leser so überraschend kommen, wie es mir im Jahre 1909 überraschend kam, als ich es aus einer Arbeit von Thue gelernt hatte: Jede diophantische Gleichung
>
> $$a_n x^n + a_{n-1} x^{n-1} y + \cdots + a_0 y^n = a,$$
>
> wo $n \geq 3$, hat, wenn die Form links nicht in homogene Faktoren niedrigeren Grades mit ganzzahligen Koeffizienten zerlegt werden kann, nur endlich viele Lösungen.
>
> E. Landau [52], p. 64

5.1. Thue's Theorem

Let α be an algebraic number of degree n. For integers p and q with $q > 0$, Thue's theorem of 1908 [122], pp. 284-305, asserts that

$$(5.1.1) \qquad |\alpha - \frac{p}{q}| > c/q^{\kappa} \qquad [c = c(\alpha,\kappa) > 0],$$

provided $\kappa > n/2 + 1$. Note that merely the existence of a positive c (depending only on α and κ) is asserted. Thus it follows that (1.3) has only finitely many solutions, since (1.5) is valid, but it does not yet follow that we can decide whether or not (1.3) actually has a solution. In fact, for $\kappa = n/2 + 1.01$ it is an open question (1971) whether or not $c = c(\alpha)$ is effectively computable.

Inequality (5.1.1) is actually true for $\kappa > 2$ (Roth's theorem), but the proof of this is more abstruse and will be deferred to Chapters 6 and 7.

This chapter is based on Davenport [23]; he attributes the use of Gauss's lemma in 5.4 to T. Schneider. Here $C_1, C_2, \ldots$ denote positive constants that depend effectively on α. The reader may take $C_1 = 2$ and $C_5 = 1$.

5.2. The Idea of Thue's Proof

Call p/q a good approximation to the real algebraic number α if $|\alpha - p/q| < 1/q^{\kappa}$, where κ is "relatively large." If there were infinitely many good approximations to α, how could we find some of them? Consider, for example, $\alpha = 2^{1/3}$. This α is a root of $f(x) = x$, where

$$f(x) = 6x/(x^3 + 4), \qquad (5.2.1)$$

so if p/q is good, we might expect that $f(p/q)$ is "almost" good. The continued fraction expansion of α yields the approximations p/q of the following table:

p/q	1	4/3	5/4	29/23	34/27
$f(p/q)$	6/5	54/43	160/127	92,046/73,057	37,179/29,509

Define $\kappa = \kappa(p/q)$ by $|\alpha - p/q| = q^{-\kappa}$. The values of κ corresponding to the entries in the preceding table are as follows:

*	2.37...	3.32...	2.21...	2.22...
1.74...	1.46...	1.95...	1.27...	1.44...

That $f(p/q)$ is at least closer to α than p/q follows from the identity

$$\alpha(x^3 + 4) - 6x = (x - \alpha)^2(\alpha x + 2\alpha^2). \tag{5.2.2}$$

In general, if α is algebraic of degree n, one can hope to find a nontrivial identity of the form

$$P(x) - \alpha Q(x) = (x - \alpha)^h[F_0(x) + \alpha F_1(x) + \cdots \alpha^{n-1}F_n(x)], \tag{5.2.3}$$

where $d(F_i(x)) \leq k$ (say), $i = 0,\ldots,n$, and hence $d(P),d(Q) \leq h + k$. Now set $f(x) = P(x)/Q(x)$. If p/q were a good approximation to α, then $f(p/q)$ would at least be closer to α (for q large) and hopefully "almost" good. Moreover, from one identity of type (5.2.3), we can find others; differentiate with respect to x!

Thue began his proof by showing (essentially by Lemma 2.2.1) that such identities exist with h arbitrarily large and with the coefficients of the $F_i(x)$ not "too large" (i.e., of at most exponential growth in h). Other authors have constructed better identities for restricted classes of α by means of hypergeometric series (see Baker [2],[3], Mahler [64], and Siegel [109]).

Thue observed that if p_1/q_1, p/q are distinct good approximations to α, with $q > q_1$, then q/q_1 is large and grows rapidly with q. In other words, the set of integers q that are denominators of good approximations (the set of good denominators) is very thin. For example, if

$$\left|\alpha - \frac{p}{q}\right| < q^{-2-\varepsilon} \qquad [(p,q) = 1]$$

for some fixed $\epsilon > 0$ and the same holds for p_1/q_1, where $q_1 < q$, then

$$q/q_1 > q_1^{\epsilon-\eta} \tag{5.2.4}$$

for any $\eta > 0$ when q_1 is sufficiently large. This follows from

$$\begin{aligned} 1 \le |pq_1 - p_1 q| &\le q_1|q - \alpha q| + q|p_1 - \alpha q_1| \\ &\le q_1 q(q_1^{-2-\epsilon} + q^{-2-\epsilon}) \end{aligned} \tag{5.2.5}$$

since if (5.2.4) were false, the right-hand side of (5.2.5) would tend to zero. We state this observation as a general principle:

(5.2.6) Good (or "almost" good) approximations repel one another; that is, the good denominators are far apart.

Now the idea of Thue's proof can be explained. If there are many good approximations to α, identities of type (5.2.3) will produce so many "almost" good approximations that by (5.2.6) there can be no further good ones.

5.3. Existence of the Identity

Assume that α is an algebraic integer [see Exercise 5.7.2(a)]. Write

$$\Phi = (x - \alpha)^h \sum_{i=0}^{n-1} \alpha^i F_i(x) = \sum_{j=0}^{h+k} \sum_{i=0}^{n-1} y_{ij}(x_1,\ldots,x_N)\alpha^i x^j, \tag{5.3.1}$$

where $\{x_1,\ldots,x_N\}$ is the set of all coefficients of the polynomials $F_i(x)$. Each y_{ij} is linear in $x_1,\ldots,x_N$, and (5.2.3) is equivalent to the $M = (n - 2)(h + k + 1)$ equations

$$y_{ij}(x_1,\ldots,x_N) = 0 \qquad (2 \le i \le n-1,\ 0 \le j \le h + k)$$

in $N = n(k + 1)$ variables. Temporarily consider α as a variable and expand Φ; clearly the numerical coefficient of each α^i for $0 \le i \le h + n - 1$ is at most 2^h. Next expand each α^i with $n \le i \le h + n - 1$ in terms of $1,\alpha,\ldots,\alpha^{n-1}$. It follows from Lemma 4.1.4 that the resulting numerical coefficients of the α^i, $0 \le i \le n - 1$, are at most

$$H_0 = 2^h[1 + (H + 1) + \cdots + (H + 1)^h] \qquad [H = H(\alpha)].$$

Lemma 5.3.1. Assume that

$$\text{(5.3.2)} \qquad \frac{1}{2}(n - 2)h - 1 < k < C_1^h - 1 \qquad (1 < C_1).$$

Then there is a nontrivial identity of the type (5.2.3) in which

$$\text{(5.3.3)} \qquad \log|x_i| \le C_2(\alpha) \frac{(n - 2)h(h + k + 1)}{n(k + 1) - (n - 2)(h + k + 1)} \qquad (1 \le i \le N),$$

where $x_1,\ldots,x_N$ are all coefficients of the polynomials $F_j(x)$, $0 \le j \le n - 1$.

Proof. The left-hand side of (5.3.2) asserts that $M < N$, so the result follows from Lemma 2.2.1. (Here the H of Lemma 2.2.1 is H_0, and (5.3.3) holds with $C_2(\alpha) = \log\{C_1 n(2H(\alpha) + 2)^2\}$.) □

Note that if k exceeds the left-hand side of (5.3.2) only slightly (e.g., $k = [(\frac{1}{2}(n - 2)h - 1)(1 + \delta)]$ for some constant $\delta > 0$), then (5.3.3) yields

(5.3.4) $$|x_i| < c_3^h \qquad [0 < c_3 = c_3(\alpha)],$$

and hence

(5.3.5) $$H(P), H(Q) < c_4^h \qquad [0 < c_4 = c_4(\alpha)].$$

5.4. A Good Approximation of α Generates Additional (Almost) Good Approximations Much Closer to α

Assume that

(5.4.1) $$\left|\alpha - \frac{p}{q}\right| < c_5/q^{\kappa} \qquad (0 < c_5)$$

and that the same holds for p_1/q_1, $q < q_1$. We shall use the nontrivial identity (5.2.3) to generate almost good approximations to α from p/q and hence obtain further restrictions on the denominator of p_1/q_1. Some candidates for almost good approximations are P_ν/Q_ν, $\nu = 0,1,2,\ldots$, where

(5.4.2) $$P_\nu = P^{(\nu)}(p/q)$$

and

(5.4.3) $$Q_\nu = Q^{(\nu)}(p/q),$$

since if ν is fixed and k is so chosen that (5.3.4) is valid, then by (5.2.3) we have $P_\nu - \alpha Q_\nu \to 0$ as $h \to \infty$. There are two difficulties here. First, all the P_ν/Q_ν might be equal; second, it might happen that $Q_\nu = 0$ for some ν.

If $Q(x) \equiv 0$, then $(x - \alpha)^h | P(x)$, and hence (see Exercise 5.7.3) $nh \le d(P) \le h + k$. But we can postulate that $k < (n - 1)h$ (see the end of 5.3) and thus exclude this possibility. Now write $D = d/dx$.

By (5.2.3),

$$(5.4.4)\qquad -Q^2(x)\ D\left[\frac{P(x)}{Q(x)}\right] = Q'(x)P(x) - Q(x)P'(x) = (x - \alpha)^{h-1}W(x)$$

for some polynomial $W(x)$, and moreover $W(x) \not\equiv 0$. Otherwise $P(x) = cQ(x)$ for some constant c, and $c \neq \alpha$ since (5.2.3) is nontrivial. Thus $(x - \alpha)^h | Q(x)$ and $nh \le d(Q) \le h + k$, a possibility we have already excluded.

If $W(p/q) \neq 0$, then $Q_1P_0 - Q_0P_1 \neq 0$; otherwise $(x - p/q)^t \| W(x)$ for some $t \ge 1$. Write

$$W(x) = (x - \frac{p}{q})^t W^*(x) = (qx - p)^t\ \frac{W^*(x)}{q^t} \qquad [(p,q) = 1].$$

Let w be the lead coefficient of W(x). Then w is also the lead coefficient of $W^*(x)$, and by Gauss's lemma (Theorem 4.1.1), $q^t | w$. Hence by the comment at the end of Section 5.3,

$$(5.4.5)\qquad q^t \le |w| \le 2(h + k)C_4^{2h} < C_6^h.$$

Since $D^tW(x) \neq 0$ at $x = p/q$,

$$(5.4.6)\qquad D^t[Q'(x)P(x) - Q(x)P'(x)] \neq 0$$

at $x = p/q$. If the left-hand side of (5.4.6) is expanded into a sum, one of the terms must be nonzero. Hence there are indices $\nu_1 \neq \nu_2$ with $0 \le \nu_1, \nu_2 \le t + 1$ such that

$$(5.4.7)\qquad Q_{\nu_2}P_{\nu_1} \neq Q_{\nu_1}P_{\nu_2}.$$

5.5. Application of the General Principle (5.2.6)

Without loss of generality, we can assume that either $Q_\nu = Q_{\nu_1} = 0$ or $P_\nu/Q_\nu = P_{\nu_1}/Q_{\nu_1} \neq p_1/q_1$. Define

$$A = q^{h+k-\nu}P_\nu/\nu!; \qquad B = q^{h+k-\nu}Q_\nu/\nu!. \tag{5.5.1}$$

By (5.4.1) and the hint to Exercise 4.2.1(c),

$$|P_\nu| \leq \nu!2^{2(h+k)}H(P(x))\max(1, |p/q|^{h+k})$$
$$\leq \nu!2^{2(h+k)}C_4^h(1 + C_5 + |\alpha|)^{h+k} \leq \nu!C_7^h$$

and similarly for Q_ν, so

$$|A|, |B| \leq q^{h+k-\nu}C_7^h. \tag{5.5.2}$$

Also, by (5.4.5) we have $\nu \leq t + 1 \leq h(\log C_6)/(\log q) + 1$, so $\nu/h \to 0$ as $h,q \to \infty$.

Now, since A and B are both integers and $AB \neq 0$,

$$1 \leq |p_1B - q_1A| \leq |B||p_1 - \alpha q_1| + q_1|A - \alpha B| \tag{5.5.3}$$
$$\leq \frac{q^{h+k-\nu}C_7^hC_5}{q_1^{\kappa-1}} + \frac{q_1q^{h+k-\nu}|P^{(\nu)}(p/q) - \alpha Q^{(\nu)}(p/q)|}{\nu!}$$
$$\leq \frac{q^{h+k-\nu}C_7^hC_5}{q_1^{\kappa-1}} + q_1q^{h+k-\nu}C_8^h(1 + C_5 + |\alpha|)^{h+k}\left|\alpha - \frac{p}{q}\right|^{h-\nu};$$

for an explicit calculation of C_8 see Exercise 5.7.4. Thus

$$1 \leq C_9^h(q^{h+k-\nu}q_1^{-\kappa+1} + q^{h+k-\nu-(h-\nu)\kappa}q_1). \tag{5.5.4}$$

Formula (5.5.4) is the decisive inequality. It says that for each h there is an interval $I(h,q)$ that q_1 must avoid. We shall show that if κ is large enough, these intervals overlap.

5.6. The Overlapping of the Intervals

Let

$$(5.6.1)\quad I(h,q) = [C_{10}^{h} q^{(h+k-\nu)/(\kappa-1)}, C_{10}^{-h} q^{(h-\nu)\kappa-(h+k-\nu)}],$$

where $C_{10} > 2C_9$. Then (assume $\kappa > 2$) it easily follows from (5.5.4) that $q_1 \notin I(h,q)$. We now show that, for h and q sufficiently large, the intervals $I(h,q)$, $h = 1,2,\ldots$, are (i) nonempty, (ii) overlapping, and (iii) tending toward infinity. For this purpose we assume that $k/h \to (n - 2)(1 + \delta)/2$ as $h \to \infty$ for some fixed $\delta > 0$; in fact, we could have defined k by $k \leq (n - 2)(1 + \delta)h/2 < k + 1$ throughout.

Assertion (iii) follows from

$$h \log C_{10} + \frac{(h + k - \nu)(\log q)}{(\kappa - 1)} \to \infty$$

(assume $C_9 > 1/2$). Both (i) and (ii) would follow from

$$(5.6.2)\quad \frac{(2h + 1)(\log C_{10})}{(\log q)} + \frac{h + 1 + k - \nu}{\kappa - 1} < (h - \nu)\kappa - (h + k - \nu)$$

for h,q sufficiently large. On dividing through by h, this is seen to be equivalent to

$$(5.6.3)\quad (1 + k/h)(\kappa - 1)^{-1} < \kappa - 1 - k/h,$$

which is satisfied by $\kappa > \frac{n}{2} + 1$ if δ is small enough.

5.7. The Proof

Assume $\kappa > \frac{n}{2} + 1$ and that there are an infinite number of fractions p/q satisfying (5.4.1). Then there are pairs p/q, p/q_1 satisfying (5.4.1) with $q < q_1$ and q_1 arbitrarily large. The intervals $I(h,q)$, $h = 1,2,\ldots$ depend only on q, and hence so does $s(q) = \sup\{\underset{\sim}{R} - \cup_{h=1}^{\infty} I(h,q)\}$. But in Section 5.6 we saw that, for q sufficiently large, $s(q) < \infty$, so q_1 cannot be arbitrarily large, a contradiction.

Formula (5.1.1) now follows from the fact that (5.4.1) has only finitely many solutions [Exercise 5.7.2(b)]. □

The following remark may be omitted on a first reading. The preceding proof is said to be nonconstructive in that it does not provide any way of actually calculating the $c = c(\alpha,\kappa)$ of (5.1.1). Indeed, if we fix a $\kappa > \frac{n}{2} + 1$, we can find C_{11}, C_{12} such that for $h_0 > C_{12}$, whenever

$$(5.7.1) \qquad C_{11} < q \quad \text{and} \quad \left|\alpha - \frac{p}{q}\right| < C_5 q^{-\kappa}$$

then

$$(5.7.2) \qquad \cup_{h \geq h_0} I(h,q) = [C_{10}^{h_0} q^{(h_0+k_0-\nu_0)/(\kappa-1)}, \infty) = [s_0(q), \infty).$$

Hence if p_1/q_1 satisfies (5.4.1), we have

$$(5.7.3) \qquad q_1 \leq \begin{cases} s_0(q) & \text{if (5.7.1) holds for some } p/q, \\ C_{11} & \text{otherwise.} \end{cases}$$

Moreover, from an effective upper bound for q_1 we could effectively compute some c for which (5.1.1) would hold [Exercise 5.7.2(b)]. But (5.7.3) is not necessarily effective since we cannot determine whether or not a fraction p/q satisfying (5.7.1) exists (1971). [If (5.7.1) held for some class of α's, we could find q by a direct search and hence effectively compute some c for (5.1.1), although this c might not be explicitly bounded below.]

Exercises

5.7.1. For $m,n \in \mathbb{Z}$ with $0 \leq m \leq n$ obtain the estimate $\binom{n}{m} \leq 2^n$ and show it is not best possible.

5.7.2. (a) Given that (5.1.1) is valid for algebraic integers α, show that it is valid for any algebraic α.

(b) For $d(\alpha) \geq 0$ and any fixed $C_5 > 0$ show that if (5.4.1) has only finitely many solutions, then (5.1.1) holds for some $c = c(\alpha, \kappa)$. Also show that if there is an effective upper bound for the largest q for which (5.4.1) is valid, then the $c = c(\alpha, \kappa)$ is effectively computable. (Hint: In (a), (5.1.1) will hold with $c = c(a_n\alpha, \kappa)/a_n$, where $P(x) = a_n x^n + \cdots + a_0$ is the minimal polynomial of α. For (b), consider the minimal value of $|\alpha - p/q|q^{\kappa}$.)

5.7.3. If $P(x) \in \mathbb{Z}[x]$, $d(\alpha) = n$, and $(x - \alpha)^h | P(x)$, show that $d(P) \geq nh$. (Hint: Let Q(x) be the minimal polynomial of α. Show that $Q(x) | P(x)$ and $(x - \alpha)^{h-1} | P(x)/Q(x)$.)

5.7.4. If $R(x) = (x - \alpha)^h S(x)$ and $\nu \leq h$, then $R^{(\nu)}(x) = (x - \alpha)^{h-\nu} S_\nu(x)$ [set $S_0(x) = S(x)$]. Estimate $H(S_\nu)$ in terms of H(S). (Hint: By Leibniz's formula,

$$R^{(\nu)}(x) = \sum_{\mu=0}^{\nu} \binom{\nu}{\mu}[(x - \alpha)^h]^{(\nu-\mu)} S^{(\mu)}(x)$$

$$= (x - \alpha)^{h-\nu} \sum_{\mu=0}^{\nu} \binom{\nu}{\mu} \frac{h!}{[h - (\nu - \mu)]!} (x - \alpha)^{\mu} S^{(\mu)}(x).$$

Now $L(S^{(\mu)}(x)) \leq \mu!\binom{d(S)+1}{\mu+1}H(S)$, so

$$H(S_\nu) \leq L(S_\nu) \leq \sum_{\mu=0}^{\nu} \binom{\nu}{\mu} \frac{h!}{[h - (\nu - \mu)]!} (1 + |\alpha|)^{\mu} \mu! \binom{d(S)+1}{\mu+1} H(S)$$

$$= \nu! H(S) \sum_{\mu=0}^{\nu} \binom{h}{\nu-\mu}(1 + |\alpha|)^{\mu}\binom{d(S)+1}{\mu+1}$$

$$\leq \nu!(1 + |\alpha|)^{\nu} 2^{h+d(S)+1} H(S).)$$

5.7.5. In the notation of Thue's theorem, show that for $s,t > 0$ there is a computable $C_{13} = C_{13}(\alpha,s,t)$ such that if

$$|\alpha - \frac{p}{q}| < q^{-\frac{n}{2}-1-s} \qquad (C_{13} < q),$$

then there is a computable $C_{14} = C_{14}(\alpha,s,t,q)$ such that

$$|\alpha - \frac{p}{q}| < q^{-\frac{n}{2(s+1)}-1-t}, \qquad (C_{14} < q)$$

has no solutions. (Hint: The only important restrictions on h,k in this chapter are $n/2 - 1 < k/h < n - 1$. Moreover, the reasoning of Section 5.6 shows that a sufficiently large solution to $|\alpha - p/q| < q^{-\kappa_1}$ will prevent $|\alpha - p/q| < q^{-\kappa_2}$ from having infinitely many solutions, provided $k/h < k_1 - 1 - \kappa_1/\kappa_2$. The case $s = n/2 - 1$ yields a result very similar to Roth's theorem (Theorem 6.1.1). See Thue [122], p. 301, Theorem III.)

*5.7.6. Find some $\kappa = \kappa(n) < n$ for which an integer q can be effectively computed such that (5.4.1) with $C_5 = 1$ has at most one solution p_1/q_1 with $q_1 > q$. (Hint: See Davenport [23] and Schinzel [96].)

Chapter 6

ROTH'S LEMMA

6.1. Historical Background of Roth's Theorem

The assumption

$$\kappa > \frac{1}{2}n + 1 \tag{6.1.1}$$

in Thue's theorem is too strong. Siegel [107] replaced (6.1.1) by $\kappa > 2\sqrt{n}$, and then in 1947 F. J. Dyson and (independently) A. O. Gelfond replaced this by $\kappa > \sqrt{2n}$. Dyson remarked that further results in this direction would probably require the use of polynomials in more than two variables. The history of these early improvements has been outlined by Dyson [30] and also by Gelfond [39], pp. 1-10.

In 1958 K. F. Roth received a Fields medal for the following theorem:

Theorem 6.1.1 (Roth [93]). Let $\kappa > 2$. If α is algebraic and $d(\alpha) \geq 3$, then there are only finitely many pairs of integers p,q with $q > 0$ such that

$$|\alpha - \frac{p}{q}| < 1/q^{\kappa}. \tag{6.1.2}$$

Equivalently, for $\epsilon > 0$ there is some $c = c(\alpha,\epsilon) > 0$ such that

$$|\alpha - \frac{p}{q}| > c/q^{2+\epsilon}. \tag{6.1.3}$$

Note that the lower bound for κ is <u>independent</u> of n.

In 1969 this was generalized by E. Wirsing as follows:

<u>Theorem 6.1.2 (Wirsing [59])</u>. Let α be algebraic. Then there are only finitely many algebraic β with $d(\beta) = n$ and $H(\beta) = H$ such that

$$(6.1.4) \qquad |\alpha - \beta| < 1/H^{2n+\epsilon}.$$

Theorem 6.1.2 was announced by Roth in 1955, but the proof was valid only for n = 1. Finally, in 1970 W. Schmidt replaced (6.1.4) by

$$(6.1.5) \qquad |\alpha - \beta| < 1/H^{n+1+\epsilon},$$

and here the exponent is best possible as a function of n (see Schmidt [99], [100], and [101]). Schmidt's proof is rather intricate, so we shall be content to obtain Theorem 6.1.2 in this volume.

Two still unanswered questions are (1) whether the c of (6.1.3) is effectively computable and (2) to what extent the ϵ of (6.1.3) can be replaced by a function $f = f(\alpha,\epsilon,q)$ with $f \to 0$ as $q \to \infty$. The first has already been discussed in connection with Thue's theorem and will be pursued further when we reach A. Baker's work in Chapters 9 through 12 (see Exercise 12.4.2). The second question has been posed by K. Mahler as follows: Is there some $c = c(\alpha,\epsilon) > 0$ such that

$$(6.1.6) \qquad |\alpha - \frac{p}{q}| > 1/q^{2+c/\log\log q}\ ?$$

A partial result in this direction has been obtained by Cugiani [21] (see Exercise 7.5.5).

6.2. The Nature of Roth's Proof

In the *Proceedings of the International Congress of Mathematicians 1958*, Roth ([94], pp. 203-210) outlined his proof as follows: Suppose $p_1/q_1,\ldots,p_m/q_m$ all satisfy $|\alpha - p_i/q_i| < 1/q_i^{\kappa}$. For any polynomial $P(x_1,\ldots,x_m) \in \mathbb{Z}[x_1,\ldots,x_m]$ with $d_{x_i}(P) = r_i$,

$$|P(\frac{p_1}{q_1},\ldots,\frac{p_m}{q_m})| \geq \frac{1}{q_1^{r_1}\cdots q_m^{r_m}}$$

provided $P \neq 0$. Let the Taylor expansion of P about $(\alpha,\ldots,\alpha)$ be

$$P(x_1,\ldots,x_m) = \Sigma\, C_{i_1\cdots i_m}(x_1 - \alpha)^{i_1}\cdots(x_m - \alpha)^{i_m},$$

where $C_{i_1\cdots i_m} \in \mathbb{Q}(\alpha)$. The (usual) idea is that we can get a contradiction if this is too small at $(p_1/q_1,\ldots,p_m/q_m)$. This will happen if $C_{i_1\cdots i_m} = 0$ when the i_j are small and if $\Sigma|C_{i_1\cdots i_m}|$ is small. Roth found a polynomial P such that

$$\Sigma|C_{i_1\cdots i_m}| \leq (q_1^{r_1}\cdots q_m^{r_m})^{\delta} \tag{6.2.1}$$

and

$$C_{i_1\cdots i_m} = 0 \quad \text{if} \quad q_1^{i_1}\cdots q_m^{i_m} \leq (q_1^{r_1}\cdots q_m^{r_m})^{\phi} \tag{6.2.2}$$

with δ "small" and ϕ "relatively large." Thus

$$|P(\frac{p_1}{q_1},\ldots,\frac{p_m}{q_m})| \le (q_1^{r_1}\cdots q_m^{r_m})^{\delta}\max(q_1^{i_1}\cdots q_m^{i_m})^{-\kappa}$$

$$\le (q_1^{r_1}\cdots q_m^{r_m})^{\delta-\kappa\phi}.$$

Hence we have a contradiction if $\delta - \kappa\phi < -1$, that is, $\kappa > (1 + \delta)/\phi$. To do this, Roth used the pigeonhole principle to obtain $C_{i_1\cdots i_m}$ for which δ became arbitrarily small and ϕ arbitrarily close to 1/2. The older work had m = 2 and hence fewer $C_{\cdots}$, so ϕ was required to be much smaller $[\phi \cong d(\alpha)^{-1/2}$ at best].

Of course, all this works only if

$$P(\frac{p_1}{q_1},\ldots,\frac{p_m}{q_m}) \ne 0, \tag{6.2.3}$$

and choosing P so that this is true is the <u>main</u> <u>difficulty</u>.

Roth overcame this problem by means of Roth's lemma, which (very roughly) asserts that if the q_i increase rapidly and the r_i decrease rapidly ($i = 1,\ldots,m$), then P cannot have a high-order zero at $(p_1/q_1,\ldots,p_m/q_m)$. Thus we want to choose a rather <u>asymmetric</u> polynomial. The proof of this lemma is by induction on m.

6.3. Generalized Wronskians

Given a polynomial $P(x_1,\ldots,x_m) \ne 0$, there do not necessarily exist polynomials $Q_1(x_1,\ldots,x_{m-1})$, $Q_2(x_m)$ such that $P = Q_1Q_2$. However, $W_P(x_1,\ldots,x_m) \ne 0$ does have this property, W_P being a certain generalized Wronskian corresponding to P. Thus an induction hypothesis on m will give us information on the factors of W_P, hence on W_P, and hence (hopefully) on P itself.

Lemma 6.3.1. Let $f_j(x) \in \underset{\sim}{Q}(x)$, $j = 1,\ldots,n$. Then the f_j are linearly independent if and only if

$$\det(\Delta_i f_j) \neq 0, \tag{6.3.1}$$

where $\Delta_i = d^{i-1}/dx^{i-1}$, $i = 1,\ldots,n$.

Proof. The linear independence obviously follows from (6.3.1). Next, assume that the f_j are linearly independent. Then $f_1 \neq 0$, and the lemma is clearly true for $n = 1$. Assume the lemma is true for $n - 1$ and proceed by induction. It suffices to show $\mathrm{Det} \neq 0$, where

$$\mathrm{Det} = \det[\Delta_i(f_j/f_1)] = f_1^{-n}\det(\Delta_i f_j) \tag{6.3.2}$$

(see Exercise 6.3.2). But

$$\mathrm{Det} = \begin{vmatrix} \Delta_1 1 & \Delta_1(f_2/f_1) & \cdots & \Delta_1(f_n/f_1) \\ \Delta_2 1 & \Delta_2(f_2/f_1) & & \Delta_2(f_n/f_1) \\ \vdots & \vdots & & \vdots \\ \Delta_n 1 & \Delta_n(f_2/f_1) & \cdots & \Delta_n(f_n/f_1) \end{vmatrix} \tag{6.3.3}$$

and $\Delta_2(f_2/f_1),\ldots,\Delta_2(f_n/f_1)$ are linearly independent; otherwise

$$\Delta_2[c_2(f_2/f_1) + \cdots + c_n(f_n/f_1)] = 0 \tag{6.3.4}$$

for constants c_i not all zero, $2 \leq i \leq n$, and hence $c_2f_2 + \cdots + c_nf_n = c_1f_1$ for some further constant c_1. Hence by the induction hypothesis the lower right $(n - 1) \times (n - 1)$ subdeterminant in (6.3.3) does not vanish, and since $\Delta_1 1 = 1$, $\Delta_2 1 = \cdots = \Delta_n 1 = 0$, the result follows. □

<u>Definition 6.3.1</u>. Define differential operators

$$\Delta_i = \Delta_i(i_1, \ldots, i_m) = \frac{\partial^{i_1 + \cdots + i_m}}{\partial x_1^{i_1} \cdots \partial x_m^{i_m}},$$

where the degree of $\Delta_i(i_1, \ldots, i_m)$ satisfies

$$\deg \Delta_i = i_1 + \cdots + i_m \leq i - 1.$$

(This definition is temporary; see Definition 6.3.2.)

Lemma 6.3.1 is the case $m = 1$ of the following lemma:

<u>Lemma 6.3.2 (Ostrowski [80])</u>. Let $f_j(x_1, \ldots, x_m) \in \underset{\sim}{Q}(x_1, \ldots, x_m)$, $j = 1, \ldots, n$. Then the f_j are linearly independent if and only if there are Δ_i, $i = 1, \ldots, n$, such that

(6.3.5) $$\det(\Delta_i f_j) \neq 0.$$

<u>Proof</u>. The linear independence obviously follows from (6.3.5). Next, assume that the f_j are linearly independent. We shall use a double induction. When $m = 1$, the lemma is true for any n by Lemma 6.3.1. Assume now that it is true for $m - 1$ and any n; we shall show that it is true for m and any n. Since $f_1 \neq 0$, it is true for $n = 1$; we proceed by induction on n. It suffices to show $\mathrm{Det} \neq 0$, where

(6.3.6) $$\mathrm{Det} = \det[\Delta_i(f_j/f_1)] = \sum_{k=1}^{K} a_k(x_1,\ldots,x_m)\det(\Delta_{i,k}f_j),$$

$$a_k(x_1,\ldots,x_m) \in \underset{\sim}{Q}(x_1,\ldots,x_m),\ \deg \Delta_{i,k} \le \deg \Delta_i,$$

since this implies that at least one of the terms on the right-hand side of (6.3.6) is nonzero. To do this renumber the rational functions $1, f_2/f_1, \ldots, f_n/f_1$ as $f_1 = 1, f_2, \ldots, f_n$ in such a way that the functions of $A = \{f_1, \ldots, f_r\}$ are independent of x_1, but those of $B = \{f_{r+1}, \ldots, f_n\}$ are not. Clearly $r \ge 1$. If B is empty, we have only $(m - 1)$ variables, and the result follows; otherwise $1 \le r < n$. Now it suffices to show $\mathrm{Det}' \ne 0$, where $\mathrm{Det}' = \det(\Delta_i g_k)$ and

(6.3.7) $$g_k = \sum_{j=1}^{n} c_{kj} f_j \qquad (c_{kj} \in \underset{\sim}{Q};\ k = 1,\ldots,n)$$

since

(6.3.8) $$\mathrm{Det}' = \det\left(\sum_{j=1}^{n} c_{kj}\Delta_i f_j\right) = \det(c_{kj})\det(\Delta_i f_j).$$

If

(6.3.9) $$f^*_{r+1} = c_{r+1}f_{r+1} + \cdots + c_n f_n$$

is independent of x_1 for constants $c_{r+1}, \ldots, c_n$ not all zero (we may assume that this occurs with $c_{r+1} = 1$), replace A by $\{f_1, \ldots, f_r, f^*_{r+1}\}$ and B by $\{f_{r+2}, \ldots, f_n\}$. Proceeding in this way we ultimately obtain an s, $1 \le s < n$, and a set of linearly independent functions g_j satisfying (6.3.7) such that

$$g_1, \ldots, g_s$$

are independent of x_1, whereas

$$\partial g_{s+1}/\partial x_1, \ldots, \partial g_n/\partial x_1$$

are linearly independent. Hence by the induction hypothesis on n we can choose differential operators $D_1, \ldots, D_s$ so that $\deg D_i \leq i - 1$, $\det(D_i g_j) \neq 0$, $i,j = 1,\ldots,s$ and also operators $E_{s+1}, \ldots, E_n$ so that $\deg E_i \leq i - s - 1$, and $\det(E_i \partial g_j/\partial x_1) \neq 0$, $i,j = s+1,\ldots,n$. Define

$$\Delta_i = \begin{cases} D_i & \text{for} \quad 1 \leq i \leq s \\ (\partial/\partial x_1)E_i & \text{for} \quad s+1 \leq i \leq n . \end{cases}$$

Note that $\deg(\partial E_i/\partial x_1) \leq 1 + i - s - 1 = i - s \leq i - 1$. Hence

(6.3.10)
$$\det(\Delta_i g_j) = \frac{\begin{vmatrix} \Delta_1 g_1 & \cdots & \Delta_1 g_s & \Delta_1 g_{s+1} & \cdots & \Delta_1 g_n \\ \vdots & & \vdots & \vdots & & \vdots \\ \Delta_s g_1 & \cdots & \Delta_s g_s & \Delta_s g_{s+1} & \cdots & \Delta_s g_n \end{vmatrix}}{\begin{vmatrix} \Delta_{s+1} g_1 & \cdots & \Delta_{s+1} g_s & \Delta_{s+1} g_{s+1} & \cdots & \Delta_{s+1} g_n \\ \vdots & & \vdots & \vdots & & \vdots \\ \Delta_n g_1 & \cdots & \Delta_n g_s & \Delta_n g_{s+1} & \cdots & \Delta_n g_n \end{vmatrix}}$$

$$= \det(D_i g_j)\det(E_i \ \partial g_j/\partial x_1) \neq 0$$

since the entries of the lower left $(n - s) \times (n - s)$ subdeterminant are all zeros. □

Lemma 6.3.3. Let $P = P(x_1,\ldots,x_m) \in \underset{\sim}{Q}[x_1,\ldots,x_m]$, $P \not\equiv 0$, and set $r_i = d_{x_i}(P)$, $1 \le i \le m$. Then there is an integer n, $1 \le n \le r_m+1$, such that

$$P = \sum_{j=1}^{n} R_j(x_1,\ldots,x_{m-1})S_j(x_m), \tag{6.3.11}$$

where $R_j \in \underset{\sim}{Q}[x_1,\ldots,x_{m-1}]$, $S_j \in \underset{\sim}{Q}[x_m]$, $1 \le j \le n$, and the sets $\{R_1,\ldots,R_n\}$ and $\{S_1,\ldots,S_n\}$ are both linearly independent.

Proof. Choose n minimal so that (6.3.11) holds. Then any dependency between the R_j or S_j would allow us to replace n by n - 1, a contradiction. □

Definition 6.3.2. In the following Δ_i will always denote an operator

$$\Delta_i = \Delta_i(i_1,\ldots,i_m) = \frac{\partial^{i_1+\cdots+i_m}}{i_1!\ldots i_m!\ \partial x_1^{i_1}\cdots\partial x_m^{i_m}}$$

of degree

$$\deg \Delta_i = i_1 + \cdots + i_m \le i - 1.$$

Note that if $P \in \underset{\sim}{Z}[x_1,\ldots,x_m]$, then $\Delta_i P \in \underset{\sim}{Z}[x_1,\ldots,x_m]$. A determinant of the form $\det(\Delta_i f_j)$, $i,j = 1,\ldots,n$, is called a generalized Wronskian (of $f_1,\ldots,f_n$).

Given a polynomial $P = P(x_1,\ldots,x_m) \in \underset{\sim}{Z}[x_1,\ldots,x_m]$, write it in the form (6.3.11). Let $\Delta_1^{(R)},\ldots,\Delta_n^{(R)}$ and $\Delta_1^{(S)},\ldots,\Delta_n^{(S)}$ be some choice of operators for the R_j and S_j, respectively, such that the corresponding generalized Wronskians do not vanish; such operators exist by Lemma 6.3.2. Form $\Delta_{\mu\nu} = \Delta_\mu^{(R)}\Delta_\nu^{(S)}$, $\mu,\nu = 1,\ldots,n \le r_m + 1$. Then

$$\Delta_{\mu\nu}P = \sum_{j=1}^{n} \Delta_\mu^{(R)}R_j\Delta_\nu^{(S)}S_j,$$

so

$$\begin{aligned} W_P = W_P(x_1,\ldots,x_m) &\equiv \det(\Delta_{\mu\nu}P) \\ &= \det(\Delta_\mu^{(R)}R_j)\det(\Delta_\nu^{(S)}S_j) \\ &\equiv uv \ne 0, \end{aligned}$$

where $u \in \underset{\sim}{Q}(x_1,\ldots,x_{m-1})$ and $v \in \underset{\sim}{Q}(x_m)$. By Gauss's lemma (Theorem 4.1.1) we can find polynomials $U = U(x_1,\ldots,x_m) \in \underset{\sim}{Z}[x_1,\ldots,x_{m-1}]$ and $V = V(x_m) \in \underset{\sim}{Z}[x_m]$ such that

$$W_P = UV. \tag{6.3.12}$$

This proves the assertion made at the beginning of this section.

Exercises

6.3.1. Let $D_{ij} = (d/dx)^i(d/dy)^j$ for $i,j = 0,1,2$, let $P = P(x,y) = ax^3 + bx^2y^2 + cy^3$ $(abc \ne 0)$, and define

$$D(k) = \det(D_{ij}P)_{i,j=0}^{k}.$$

Show that $D(2) = -36abcx^2y^2$, but that there do not exist polynomials $Q_1 = Q_1(x)$ and $Q_2 = Q_2(y)$ such that $D(k) = Q_1Q_2$ for $k < 2$. (Hint: $D(1) = -xy(2abx^3 + 9acxy + 2bcy^2)$.)

6.3.2. Prove formula (6.3.2) by means of Leibniz's rule and row operations.

6.3.3. Does Lemma 6.3.1 remain valid if the $f_j(x)$ are formal power series in x?

6.3.4. Show that Roth's theorem is best possible in the sense that if α is real and irrational then there are infinitely many pairs of integers p,q with $q > 0$ such that

$$|\alpha - \frac{p}{q}| < \frac{1}{q^2} .$$

(Hint: This is really due to Dirichlet, who was the first to obtain the result of Exercise 2.1.4.)

6.4 A Non-Archimedean Valuation on $\mathbb{Q}(x_1,\dots,x_m)$

Definition 6.4.1. A real-valued function w defined on a field $\mathbb{F}$ is said to be a non-Archimedean valuation if, for all $a,b \in \mathbb{F}$,

(i) $w(0) = 0$ and $w(a) > 0$ if $a \neq 0$.

(ii) $w(ab) = w(a)w(b)$.

(iii) $w(a \pm b) \leq \max[w(a),w(b)]$.

For every tuple $(\rho_1,\dots,\rho_m)$, $\rho_i > 0$, we shall construct a non-Archimedean valuation on $\mathbb{Q}(x_1,\dots,x_m)$. For $P \in \mathbb{Q}[x_1,\dots,x_m]$, write

$$\Delta_i P = \Delta_i(i_1,\dots,i_m)P = P_{i_1 \cdots i_m}$$

and define

$$I(P) = I(P;\rho_1,\dots,\rho_m;\alpha_1,\dots,\alpha_m) \tag{6.4.1}$$

$$= \min \{\frac{i_1}{\rho_1} + \cdots + \frac{i_m}{\rho_m} | P_{i_1 \cdots i_m}(\alpha_1,\dots,\alpha_m) \neq 0\}.$$

Note that $I(P) = \infty$ if the set on the right-hand side of (6.4.1) is empty. We call $I(P)$ the <u>index</u> of P at $(\alpha_1,\ldots,\alpha_m)$ <u>relative</u> to $(\rho_1,\ldots,\rho_m)$. A Taylor series expansion about $\alpha_1,\ldots,\alpha_m$ shows that

$$\text{(6.4.2)}\qquad P(\alpha_1 + x^{1/\rho_1}x_1,\ldots,\alpha_m + x^{1/\rho_m}x_m)$$

$$= \sum_{i_1=0}^{r_1} \cdots \sum_{i_m=0}^{r_m} P_{i_1\cdots i_m}(\alpha_1,\ldots,\alpha_m)x^{i_1/\rho_1+\cdots+i_m/\rho_m}x_1^{i_1}\cdots x_m^{i_m}.$$

Thus $I(P)$ is the exponent of the lowest power of x in such an expansion. Hence

(i') $I(P) \geq 0$, $I(P) = \infty$ if and only if $P \equiv 0$.

(ii') $I(P_1P_2) = I(P_1) + I(P_2)$.

(iii') $I(P_1 \pm P_2) \geq \min[I(P_1),I(P_2)]$.

For $a \in \mathbb{Q}(x_1,\ldots,x_m)$, write

$$\text{(6.4.3)}\qquad a = P/Q \qquad (P,Q \in \mathbb{Q}[x_1,\ldots,x_m])$$

and define $w(a) = \exp(I(Q) - I(P))$. The function w is well defined since by (ii') it gives the same value for any P,Q satisfying (6.4.3). It follows immediately that w has properties (i) and (ii); to show (iii), say $a = P_1/Q_1$, $b = P_2/Q_2$, and $w(a) \leq w(b)$; that is, $I(Q_1P_2) \leq I(P_1Q_2)$. Then by (iii')

$$\begin{aligned} w(a + b) &= \exp(I(Q_1) + I(Q_2) - I(P_1Q_2 + P_2Q_1)) \\ &\leq \exp\{I(Q_1) + I(Q_2) - \min[I(P_1Q_2),I(P_2Q_1)]\} \\ &= w(b). \end{aligned}$$

Hence w is a non-Archimedean valuation.

Lemma 6.4.1. (i) $I(P) = 0$ if and only if $P(\alpha_1,\dots,\alpha_m) \neq 0$.

(ii) $I(P;s\rho_1,\dots,s\rho_m;\alpha_1,\dots,\alpha_m) = s^{-1}I(P;\rho_1,\dots,\rho_m;\alpha_1,\dots,\alpha_m)$ if $s > 0$.

(iii) $I(P;\rho_1,\dots,\rho_m;\alpha_1,\dots,\alpha_m) = I(P;\rho_1,\dots,\rho_{m-1};\alpha_1,\dots,\alpha_{m-1})$ if $\partial P/\partial x_m = 0$, and similarly for the other variables.

(iv) $\max(0, I(P) - \Sigma_{i=1}^m \ell_i/\rho_i) \leq I(P_{\ell_1\cdots\ell_m})$.

Proof. We have

$$(6.4.4)\quad I(P_{\ell_1\cdots\ell_m}) = \min\{i_1/\rho_1 + \cdots + i_m/\rho_m \mid P_{\ell_1+i_1,\dots,\ell_m+i_m}(\alpha_1,\dots,\alpha_m) \neq 0\}$$

$$= \min_{j_s\geq\ell_s}\{j_1/\rho_1 + \cdots + j_m/\rho_m - (\ell_1/\rho_1 + \cdots + \ell_m/\rho_m) \mid P_{j_1\cdots j_m}(\alpha_1,\dots,\alpha_m) \neq 0\},$$

and the right-hand side of (6.4.4) can only become smaller when the restriction $j_s \geq \ell_s$ is dropped. This proves (iv); the rest is immediate. □

In the following $I(P;\rho_1,\dots,\rho_m;\alpha_1,\dots,\alpha_m)$ will usually be considered only when $d_{x_i}(P) = r_i \leq \rho_i$.

Exercises

6.4.1. Given a fixed prime number p, every $a \in \mathbb{Q}$ can be written in the form $a = p^e r/s$, where $e,r,s \in \mathbb{Z}$ and both r and s are relatively prime to p. Define $w(0) = 0$ and $w(a) = p^{-e}$. Show that w is a non-Archimedean valuation on $\mathbb{Q}$.

6.4.2. Show that $I(P;r_1,\ldots,r_m;0,\ldots,0) = m$ if $P(x_1,\ldots,x_m) = x_1^{r_1}\cdots x_m^{r_m}$.

6.5. The Index as a Function of Height and Degree

Definition 6.5.1. For $0 < B$ and $r_j \in \underset{\sim}{Z}^+$, let

$$\mathcal{P}_m = \mathcal{P}_m(B;r_1,\ldots,r_m) \tag{6.5.1}$$

$$= \{P \in \underset{\sim}{Z}[x_1,\ldots,x_m] \mid d_{x_j}(P) \le r_j,\ H(P) \le B, \text{ and } P \not\equiv 0\}.$$

(This definition of $\mathcal{P}_m$ is temporary; see Definition 6.5.2.)

Let

$$\Theta_m = \Theta_m(B;r_1,\ldots,r_m;q_1,\ldots,q_m)$$

$$= \sup_{P,\beta_j} I(P;r_1,\ldots,r_m;\beta_1,\ldots,\beta_m)$$

$$= \sup I(P),$$

where the supremum is taken over all $P \in \mathcal{P}_m$ and all $(\beta_1,\ldots,\beta_m)$ such that $\beta_j \in \underset{\sim}{A}$ and $H(\beta_j) = q_j$ for $1 \le j \le m$.

The index $I(P;r_1,\ldots,r_m;\beta_1,\ldots,\beta_m)$ measures roughly how "flat" the polynomial P is at $(\beta_1,\ldots,\beta_m)$, that is, the "degree" to which it vanishes there. Thus Θ_m tells us how badly polynomials of bounded height and degree can vanish at an algebraic point of fixed height. We expect it to increase with B and decrease with the q_i; its behavior as a function of the r_i is less clear. Roth's great contribution was to find (in some cases) a bound for Θ_m in terms of m. For m = 1 we have a general result.

<u>Lemma 6.5.1</u>. $\Theta_1(B;q_1;r_1) \leq (\log B)/(r_1 \log q_1) + (2 \log 2)/(\log q_1)$.

<u>Proof</u>. $I(P;r_1;\beta_1) = \min\{i_1/r_1 | P_{i_1}(\beta_1) \neq 0\} = j_1/r_1$, where $(x_1 - \beta_1)^{j_1} \| P(x_1)$. Now assume that $P \in \underset{\sim}{Z}[x_1]$, $H(P) \leq B$, $d(P) \leq r_1$, and $H(\beta_1) = q_1$. Then by Gauss's lemma (Theorem 4.1.1) $P(x_1) = q(x_1)^{j_1} a(x_1)$, where $q(x_1)$ is the minimal polynomial of β_1 and $a(x_1) \in \underset{\sim}{Z}[x_1]$, so

$$(6.5.2) \qquad q_1^{j_1} = H(q(x_1))^{j_1} \leq L(q(x_1)^{j_1})L(a(x_1)) \leq 2^{r_1}L_2(P)$$

$$\leq 2^{r_1}L(P) \leq 2^{r_1}(r_1 + 1)B \leq 2^{2r_1}B$$

by Theorem 4.1.3 and (4.1.4). Thus

$$(6.5.3) \qquad I(P) = \frac{j_1}{r_1} \leq \frac{\log B}{r_1 \log q_1} + \frac{2 \log 2}{\log q_1},$$

and the result follows since the right-hand side of (6.5.3) is independent of P and β_1.

The quantity Θ_2 has been studied, but little is probably known about Θ_m for $m \geq 3$. However, by means of the technique of Section 6.3, Roth obtained a bound for Θ_m in terms of m for an infinite class of highly asymmetric polynomials.

<u>Definition 6.5.2</u>. Fix a real number t, $0 < t \leq 1$. Henceforth we consider $\vartheta_m = \vartheta_m^{(t)}$ to be defined only if

$$(6.5.4) \qquad r_{j+1} < tr_j \qquad (1 \leq j \leq m - 1).$$

Now for $P \in \mathcal{P}_m$ form $W_P = \det(\Delta_{\mu\nu}P) = U(x_1,\ldots,x_{m-1})V(x_m)$ as in Section 6.3; this is an $n \times n$ determinant with $1 \leq n \leq r_m + 1$. By (4.1.2)

$$L(W_P) \leq n! \max L(\Delta_{\mu_1 1}P \cdots \Delta_{\mu_n n}P)$$

$$\leq n!(\max L(\Delta_{\mu\nu}P))^n,$$

where

$$\Delta_{\mu\nu}P = \frac{1}{i_1!\cdots i_m!}\frac{\partial^{i_1+\cdots+i_m}}{\partial x_1^{i_1}\cdots\partial x_m^{i_m}}\sum_{j_1=1}^{r_1}\cdots\sum_{j_m=1}^{r_m} a_{j_1\cdots j_m}x_1^{j_1}\cdots x_m^{j_m}$$

$$(a_{j_1\cdots j_m} \in \mathbb{Q})$$

$$= \sum_{j_1=1}^{r_1}\cdots\sum_{j_m=1}^{r_m}\binom{j_1}{i_1}\cdots\binom{j_m}{i_m}a_{j_1\cdots j_m}x_1^{j_1-i_1}\cdots x_m^{j_m-i_m},$$

so

$$L(\Delta_{\mu\nu}P) \leq 2^{r_1+\cdots+r_m}(r_1 + 1)\cdots(r_m + 1)B \leq 2^{2(r_1+\cdots+r_m)}B$$

and

$$\text{(6.5.5)} \qquad L(W_P) \leq n!B^n 2^{2n(r_1+\cdots+r_m)} \equiv M.$$

Hence, since U,V have no variables in common, $H(W_P) \leq L(W_P) \leq M$ yields $H(U),H(V) \leq M$. Thus $U \in \mathcal{P}_{m-1}(M;nr_1,\ldots,nr_{m-1})$ and $V \in \mathcal{P}_1(M;nr_m)$. Also, by Lemma 6.4.1(iii),

$$(6.5.6)\quad I(W_P;nr_1,\ldots,nr_m;\beta_1,\ldots,\beta_m)$$

$$= I(U;nr_1,\ldots,nr_{m-1};\beta_1,\ldots,\beta_{m-1}) + I(V;nr_m;\beta_m)$$

$$\leq \Theta_{m-1}[M;nr_1,\ldots,nr_{m-1};H(\beta_1),\ldots,H(\beta_{m-1})]$$

$$+ \Theta_1[M;nr_m;H(\beta_m)]$$

$$\equiv \Phi_m.$$

The index of P will now be bounded in terms of $I(W_P)$. By (ii') and (iii') of Section 6.4 and Lemma 6.4.1(iv),

$$(6.5.7)\quad I(W_P;r_1,\ldots,r_m;\beta_1,\ldots,\beta_m) \geq \min[I(\Delta_{\mu_1 1}P\cdots\Delta_{\mu_n n}P)]$$

$$\geq \min \sum_{\nu=1}^{n} I(\Delta_{\mu_\nu \nu}P)$$

$$\geq \min \sum_{\nu=1}^{n} \max[0,I(P) - \sum_{j=1}^{m} \frac{i_j}{r_j}],$$

where I is taken with respect to $r_1,\ldots,r_m;\beta_1\ldots,\beta_m$ on the right, and

$$\Delta_{\mu_\nu \nu} = \Delta_\mu \Delta_\nu = \frac{\partial^{i_1+\cdots+i_m}}{i_1!\cdots i_m!\partial x_1^{i_1}\cdots\partial x_m^{i_m}}.$$

Since

$$i_1 + \cdots + i_{m-1} \leq \mu - 1 \leq n - 1 \leq r_m \leq tr_{m-1}$$

and

$$i_m \leq \nu - 1 \leq n - 1,$$

we have

(6.5.8) $$I(W_p;r_1,\ldots,r_m;\beta_1,\ldots,\beta_m)$$
$$\geq \sum_{\nu=1}^{n} \max[0, I(P) - \frac{\mu - 1}{r_{m-1}} - \frac{\nu - 1}{r_m}]$$
$$\geq \sum_{\nu=1}^{n} \max[0, I(P) - t - \frac{\nu - 1}{r_m}].$$

Lemma 6.5.2. Let $X, Y > 0$. If $\sum_{\nu=0}^{n-1} \max(0, Y - \nu/r) \leq X$, where $1 \leq r \in \mathbb{Z}$ and $n \leq r + 1$, then

$$Y \leq \max[2(\frac{X}{n}),\ 2(\frac{X}{n})^{1/2}].$$

Proof. Let $N = [rY] + 1$, where the brackets denote "greatest integer in."

Case (i): $n \leq N$. Then $n - 1 \leq rY$, so

$$\frac{1}{2}\, nY \leq \frac{1}{2}\, n(2Y - \frac{n - 1}{r}) = nY - \frac{n(n - 1)}{2r} \leq X.$$

Case (ii): $N < n$. Here $n/2 \leq r$ since $n \leq r + 1$. Also $rY < N \leq rY + 1$. Hence

$$\frac{nY^2}{4} \leq \frac{rY^2}{2} \leq N(Y - \frac{Y}{2}) \leq NY - \frac{N(N - 1)}{2r} \leq X.$$

□

Let $Y = I(P) - t$; then from (6.5.8), Lemma 6.4.1(ii), and (6.5.6) we obtain the following corollary:

Corollary 6.5.1. $I(P) = I(P;r_1,\ldots,r_m;\beta_1,\ldots,\beta_m) \leq t + 2\max(\Phi_m, \Phi_m^{1/2})$, where Φ_m is given by (6.5.6) and (6.5.5).

6.6. Proof of Roth's Lemma

Here the exposition is patterned after that of Mahler [67], pp. 90-97.

Definition 6.6.1. The nonnegative numbers $B;q_1,\ldots,q_m;$ $r_1,\ldots,r_m$ are said to have the property Γ_m if either $m = 1$ and

(6.6.1) $$2^{4/t} \le q_1, \quad B \le q_1^{r_1 t/2}$$

or $2 \le m$ and

(6.6.2) $$2^{m(m+1)(2m+3)/t} \le q_1, \quad B \le q_1^{r_1 t/(m+1)},$$

$r_{j+1} \le r_j t$ $(1 \le j \le m - 1)$, and $r_1 \log q_1 \le r_j \log q_j$ $(2 \le j \le m)$.

Roughly, the property Γ_m asserts that the q_j are large and increasing while the r_j are decreasing.

Lemma 6.6.1. If $2 \le m$, $1 \le n \le r_m + 1$, and $B;q_1,\ldots,q_m;r_1,\ldots,r_m$ have the property Γ_m, then [see (6.5.5)]

(6.6.3) $$M;q_1,\ldots,q_{m-1};nr_1,\ldots,nr_{m-1}$$

have the property Γ_{m-1}.

Proof. If $m = 2$, then (6.6.1) clearly holds. If $3 \le m$, only the second inequality of (6.6.2) requires verification. Since $n! \le n^n \le (r_m + 1)^n \le 2^{nt_m}$, we have

$$(6.6.4)\qquad M \le 2^{2nmr_1+nr_mB^n} \le 2^{nr_1(2m+1)} q_1^{nr_1t/(m+1)}$$

$$\le q_1^{nr_1t[1/m(m+1)+1/(m+1)]} = q_1^{(nr_1)t/m}$$

□

Lemma 6.6.2. Say $2 \le m$ and $B;q_1,\ldots,q_m;r_1,\ldots,r_m$ have the property Γ_m. Then we can find

$$(6.6.5)\qquad M;q_1,\ldots,q_{m-1};r_1',\ldots,r_{m-1}'$$

with the property Γ_{m-1} such that

$$(6.6.6)\qquad \Theta_m(B;q_1,\ldots,q_m;r_1,\ldots,r_m) \le t + 2\max(\Psi_m,\Psi_m^{1/2}),$$

where

$$(6.6.7)\qquad \Psi_m = \Theta_{m-1}(M;q_1,\ldots,q_{m-1};r_1',\ldots,r_{m-1}') + t.$$

Proof. Choose the numbers of (6.6.3) for (6.6.5); they have the property Γ_{m-1}. By Lemma 6.5.1, (6.6.4), and (6.6.2),

$$\Theta_1(M;nr_m;q_m) \le \frac{\log_2 M}{nr_m \log_2 q_m} + \frac{2}{\log_2 q_m}$$

$$\le \frac{nr_1(2m+1)}{nr_m \log_2 q_m} + \frac{nr_1 t \log_2 q_1}{(m+1)(nr_m \log_2 q_m)} + \frac{2nr_m}{nr_m \log_2 q_m}$$

$$\le \frac{r_1(2m+3)}{r_m \log_2 q_m} + \frac{t}{m+1}$$

$$\le \frac{r_1(2m+3)}{r_1 \log_2 q_1} + \frac{t}{m+1} \le \frac{t}{m(m+1)} + \frac{t}{m+1}$$

$$= \frac{t}{m} < t.$$

Hence by Corollary 6.5.1,

$$I(P;r_1,\ldots,r_m;\beta_1,\ldots,\beta_m) < t + 2\max(\Psi_m,\Psi_m^{1/2}) \tag{6.6.8}$$

if $H(\beta_j) = q_j$, $1 \le j \le m$, and (6.6.6) follows from (6.6.8) by Definition 6.5.1.

□

<u>Theorem 6.6.1 (Roth's lemma)</u>. Let $1 \le m$. If $B;q_1,\ldots,q_m;r_1,\ldots,r_m$ have the property Γ_m, then

$$\Theta_m = \Theta_m(B;r_1,\ldots,r_m;q_1,\ldots,q_m) \le c_m t^{2^{-m+1}}, \tag{6.6.9}$$

where $c_m = 2^{m+1} - 3$.

<u>Proof</u>. This is true if $m = 1$ because

$$\Theta_1(B;r_1;q_1) \le \frac{\log B}{r_1 \log q_1} + \frac{2\log 2}{\log q_1}$$

$$\le \frac{t}{2} + \frac{t}{2} = t.$$

If we assume (6.6.9) is true for $m - 1$, $2 \le m$, then by Lemma 6.6.2,

$$\Theta_m \le t + 2\max(\Psi_m,\Psi_m^{1/2}),$$

where $\Psi_m \le c_{m-1}t^{2^{-m+2}} + t$. Now $0 < t \le 1$ and $1 \le c_{m-1}$, so

$$\Theta_m \le t^{2^{-m+1}} + 2\max\{c_{m-1}t^{2^{-m+1}} + t^{2^{-m+1}},$$
$$[(c_{m-1}^2 + 2c_{m-1})t^{2^{-m+2}} + t^{2^{-m+2}}]^{1/2}\}$$

$$= t^{2^{-m+1}} + 2(c_{m-1}t^{2^{-m+1}} + t^{2^{-m+1}}) = c_m t^{2^{-m+1}}.$$

□

Note: Roth originally had $\Theta_m \le 10^m t^{2^{-m}}$.

An immediate consequence of Roth's lemma is the following theorem:

Theorem 6.6.2. Let t and B be real numbers, with $0 \le t < 1$ and $1 \le B$. Let $q_1,\ldots,q_m; r_1,\ldots,r_m$ be positive integers such that

$$(6.6.10) \qquad 2^{m(m+1)(2m+3)/t} \le q_1, \quad B \le q_1^{r_1 t/(m+1)},$$

and

$$(6.6.11) \qquad r_{j+1} \le tr_j \ (1 \le j \le m-1), \quad r_1 \log q_1 \le r_j \log q_j \quad (2 \le j \le m).$$

Let $\beta_1,\ldots,\beta_m$ be algebraic numbers of heights $q_1,\ldots,q_m$. Let $P \in \underset{\sim}{Z}[x_1,\ldots,x_m]$ with $P \not\equiv 0$, $H(P) \le B$, and $d_{x_j}(P) \le r_j$ for $1 \le j \le m$. Then there are nonnegative integers $j_1,\ldots,j_m$ such that

$$P_{j_1 \cdots j_m}(\beta_1,\ldots,\beta_m) \ne 0$$

and

$$\sum_{i=1}^{m} \frac{j_i}{r_i} \le 2^{m+1} t^{2^{-m+1}},$$

that is,

$$(6.6.12) \qquad I(P; r_1,\ldots,r_m; \beta_1,\ldots,\beta_m) \le 2^{m+1} t^{2^{-m+1}}.$$

Exercises

6.6.1. Show that if $r_{j+1} \le tr_j$ were dropped from the hypotheses of Theorem 6.6.2, we could not conclude that the index would be small. (Hint (Cassels): If $P(x_1,x_2) = (x_1 - x_2)^r$, then $I(P;r,r;1/q,1/q) = 1$.)

**<u>6.6.2</u>. Can (6.6.9) be replaced by $\Theta_m < c_1(m)t^{m^{-c_2}}$, where $0 \le c_1(m)$, $0 \le c_2$ and c_2 is independent of m? (Mahler [67], p. v; perhaps one should modify the hypothesis of Theorem 6.6.1.)

Chapter 7

WIRSING'S THEOREM

7.1. Notions of Measure Theory and Probability Theory

Since some readers may be more familiar with measure theory than with probability theory, we pause here to translate some of the notions of the latter into the former. A more detailed discussion (not necessary for our purposes) can be found on pp. 149, 150 and 170-172 of Loève [62].

Probability theory	Measure theory
Sample space $\mathscr{S}$	Set S
Random variable X	Function g on S
Probability measure P on $\mathscr{S}$	Measure μ on S, $\mu(S) = 1$
Distribution function F on X:	$\mu(x) = \mu\{s \mid g(s) \leq x\}$
$F(x) = P\{X \leq x\}$, x real	
Expectation of X:	$\nu = \int x \, d\mu(x)$
$E(x) = \nu = \int x \, dF$	
Expectation of $\phi(X)$:	$\int \phi(x) \, d\mu(x)$
$E[\phi(X)] = \int \phi(x) \, dF$	
Variance of X:	$\sigma^2 = \int (x - \nu)^2 \, d\mu(x)$
$\text{Var}(X) = \sigma^2 = \int (X - \nu)^2 \, dF$	

The integrals here are Lebesgue-Stieltjes integrals. By expanding the integral that defines Var X, one easily shows that

$$\text{Var } X = E[(X - \nu)^2] = E(X^2) - [E(X)]^2 = E(X^2) - \nu^2.$$

<u>Lemma 7.1.1 (Chebyshev's inequality)</u>.

$$P\{|X - \nu| > t\} \leq \frac{\sigma^2}{t^2}\,. \tag{7.1.1}$$

<u>Proof</u>.

$$\text{Var } X = \sigma^2 = \int (x - \nu)^2 \, d\mu(x) \geq \int_{|x-\nu|>t} (x - \nu)^2 \, d\mu(x)$$

$$\geq t^2 \lim_{\epsilon\to 0} \left[\int_{-\infty}^{\nu-t-\epsilon} d\mu(x) + \int_{\nu+t+\epsilon}^{\infty} d\mu(x) \right]$$

$$= t^2 \lim_{\epsilon\to 0} [\mu\{s \mid g(x) \leq \nu - t - \epsilon\} + \mu\{s \mid g(s) \geq \nu + t + \epsilon\}]$$

$$= t^2 \lim_{\epsilon\to 0} [\mu\{s \mid |g(s) - \nu| \geq t + \epsilon\}]$$

$$= t^2 P\{|X - \nu| > t\}.$$

□

If $X_1, \ldots, X_m$ are random variables on $\mathcal{S}$ and ϕ is a function of m variables, then $\phi(X_1, \ldots, X_m)$ is considered to be a random variable on $\mathcal{S}^m$, the Cartesian cross-product of $\mathcal{S}$ with itself m times. Thus, for example,

$$E(X_1 + X_2) = \iint (x + y) \, d\mu_1(x) \, d\mu_2(y) = E(X_1) + E(X_2)$$

since $\int d\mu_1(x) = \int d\mu_2(x) = 1$.

We are interested in the special case where

$$\mathscr{S} = \mathscr{S}(m) = \{\underline{x} = (x_1, \ldots, x_m) \mid x_i \text{ real}, \ 0 \leq x_i \leq 1\} = [0,1]^m.$$

For $0 \leq \gamma \leq 1$, define random variables on $\mathscr{S}$ by

$$X_i = X_i(\gamma, \underline{x}) = \begin{cases} 1 \text{ if } x_i \leq \gamma & (1 \leq i \leq m), \\ 0 \text{ otherwise} & (1 \leq i \leq m) \end{cases}$$

(see Fig. 3). For any i, the corresponding

$$\mu_i(x) = \mu(x) = \begin{cases} 0 & \text{if } \ x < 0, \\ 1 - \gamma & \text{if } \ 0 \leq x < 1, \\ 1 & \text{if } \ 1 \leq x, \end{cases}$$

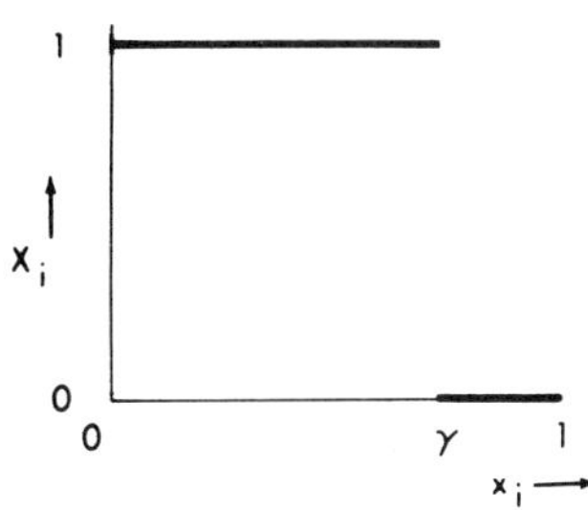

Figure 3

so

$$E(X_i) = \int x \, d\mu(x) = [(1 - \gamma) - 0] \cdot 0$$

$$+ [1 - (1 - \gamma)] \cdot 1 = \gamma$$

and

$$\sigma_i^2 = E(X_i^2) - [E(X_i)]^2 = \gamma(1 - \gamma)$$

since $X_i^2 = X_i$. For the random variable X_iX_j, $i \neq j$,

$$\mu(x) = \begin{cases} 0 & \text{if } x < 0, \\ (1 - \gamma)^2 + 2\gamma(1 - \gamma) & \text{if } 0 \leq x < 1, \\ 1 & \text{if } 1 \leq x \end{cases}$$

(see Fig. 4), so $\gamma^2 = E(X_iX_j) = E(X_i)E(X_j)$; this is also immediate from the independence of X_i and X_j.

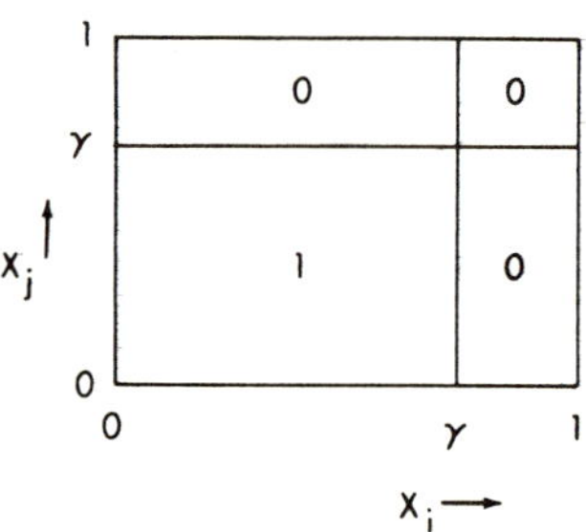

Figure 4

Definition 7.1.1. Let $S_m = S_m(\gamma,\underline{x}) = X_1 + \cdots + X_m$. One easily shows that $E(S_m) = m\gamma$ and $\text{Var}(S_m) = m\gamma(1 - \gamma)$.

Furthermore, we have the following lemma:

Lemma 7.1.2. Let μ denote Lebesgue measure in Euclidean m-space. If $T_j(m) \subseteq \mathscr{I}(m)$ and for $j = 1,\ldots,n$ we have $\mu[T_j(m)] \to 1$ as $m \to \infty$, then

$$\mu[\bigcap_{j=1}^{n} T_j(m)] \to 1 \text{ as } m \to \infty.$$

7.2 The Weak Law of Large Numbers

If $\underline{x} = (x_1,\ldots,x_m)$ is a point from $\mathscr{A}$ chosen "at random", we would expect about m/2 of its coordinates to be less than 1/2 and, more generally, about $m\gamma$ of its coordinates to be less than γ, $0 \le \gamma \le 1$.

Definition 7.2.1. For $\varepsilon > 0$, let $M_\varepsilon(m)$ be the set of all $\underline{x} = (x_1,\ldots,x_m) \in \mathscr{A}$ for which $N(\gamma,\underline{x})$, the number of coordinates of $\underline{x}$ such that $x_i \le \gamma$, satisfies

$$m(\gamma - \varepsilon) < N(\gamma,\underline{x}) < m(\gamma + \varepsilon) \tag{7.2.1}$$

for all γ with $0 \le \gamma \le 1$. We call $M_\varepsilon(m)$ the set of ε-random points (inside the m-cube).

Theorem 7.2.1 (Weak Law of Large Numbers). Let μ denote Lebesgue measure. Then

$$\mu(M_\varepsilon(m)) \to 1 \quad \text{as} \quad m \to \infty. \tag{7.2.2}$$

Proof 1. Define the random variables X_i, S_m as in Section 7.1. By Lemma 7.1.1 (Chebyshev's inequality),

$$P\{|S_m - m\gamma| > \tfrac{1}{2}\varepsilon m\} \le \frac{4\gamma(1-\gamma)}{\varepsilon^2 m} \to 0 \quad \text{as} \quad m \to \infty,$$

so

$$P\{m(\gamma - \varepsilon) < S_m < m(\gamma + \varepsilon)\} \to 1 \quad \text{as} \quad m \to \infty. \tag{7.2.3}$$

Now

$$M_\varepsilon(m) = \{\underline{x} \in \mathscr{A} \mid m(\gamma - \varepsilon) \le S_m(\gamma,\underline{x}) \le m(\gamma + \varepsilon) \text{ for all } \gamma \in [0,1]\}.$$

Fix some $\gamma_j \in [0,1]$ and set

$$T_{j,\epsilon} = T_{j,\epsilon}(m)$$
$$= \{\underline{x} \in \mathscr{S} | m(\gamma_j - \epsilon) \leq S_m(\gamma_j, \underline{x}) \leq m(\gamma_j + \epsilon)\}.$$

By (7.2.3), $\mu(T_{j,\epsilon}) \to 1$ as $m \to \infty$. Now let $\{\gamma_j\}_{j=1}^n$ be a finite set of points in $[0,1]$ such that every interval of length $\epsilon/2$ contains one of them. We shall show that

$$\cap = \bigcap_{j=1}^{n} T_{j,\epsilon} \subseteq M_{2\epsilon}(m). \tag{7.2.4}$$

For any $\gamma \in [0,1]$ pick a $\gamma_j > \gamma$ such that $|\gamma_j - \gamma| < \epsilon/2$. Then for $\underline{x} \in \cap$, since $S_m(\gamma, \underline{x})$ is nondecreasing with γ, we have

$$S_m(\gamma, \underline{x}) = \{S_m(\gamma, \underline{x}) - S_m(\gamma_j, \underline{x})\} + S_m(\gamma_j, \underline{x})$$
$$\leq 0 + m(\gamma_j + \epsilon) < m(\gamma + 2\epsilon).$$

Similarly (choose $\gamma_j < \gamma$) $m(\gamma - 2\epsilon) < S_m(\gamma, \underline{x})$, and (7.2.4) is verified. The result now follows immediately from Lemma 7.1.2.

□

<u>Proof 2</u>. Let $\underline{x} = (x_1, \ldots, x_m) \in \mathscr{S}$. The volume of an m-dimensional parallelopiped on which exactly r of the x_i satisfy $x_i \leq \gamma$ is $\gamma^r(1 - \gamma)^{m-r}$, and the number of such parallelopipeds is $\binom{m}{r}$, so

$$\mu[M_\epsilon(m)] = \Sigma' \binom{m}{r} \gamma^r (1 - \gamma)^{m-r}, \tag{7.2.5}$$

where Σ' denotes the sum over all integers r such that $m(\gamma - \epsilon) < r < m(\gamma + \epsilon)$. That $\mu[M_\epsilon(m)] \to 1$ as $m \to \infty$ can now be deduced from the DeMoivre-Laplace limit theorem, but there is a more elementary way. Since

$$\sum_{r \le m(\gamma-\varepsilon)} \binom{m}{r} \gamma^r (1-\gamma)^{m-r} = \sum_{m(1-\gamma+\varepsilon) \le m-r} \binom{m}{m-r} \gamma^r (1-\gamma)^{m-r}$$

$$= \sum_{m(1-\gamma+\varepsilon) \le r} \binom{m}{r} (1-\gamma)^r \gamma^{m-r}$$

and each individual term of (7.2.5) tends to 0 as $m \to \infty$ $(0 \le r \le m)$, it suffices to show that

$$(7.2.6) \qquad \sum_{m(\gamma+\varepsilon)<r} \binom{m}{r} \gamma^r (1-\gamma)^{m-r} = \sum_{m(\gamma+\varepsilon)<r} b(r,m,\gamma) = \Sigma_m \to 0$$

as $m \to \infty$. Now, for $m > 1/\varepsilon$,

$$(7.2.7) \qquad \frac{b(r,m,\gamma)}{b(r-1,m,\gamma)} = \frac{m-(r-1)}{r} \frac{\gamma}{1-\gamma} \le \frac{m - m\gamma - m\varepsilon + 1}{m(\gamma+\varepsilon)} \frac{\gamma}{1-\gamma}$$

$$\le \frac{\gamma}{\gamma+\varepsilon}$$

since $m(\gamma+\varepsilon) < r$. Hence Σ_m is bounded by a geometric series of ratio $\gamma/(\gamma+\varepsilon)$, so

$$\Sigma_m \le b([m(\gamma+\varepsilon)],m,\gamma) \frac{\gamma+\varepsilon}{\varepsilon}$$

$$\le b([m(\gamma+\tfrac{\varepsilon}{4})],m,\gamma) \left(\frac{\gamma}{\gamma+\varepsilon}\right)^{m\varepsilon/4} \frac{\gamma+\varepsilon}{\varepsilon}.$$

The second inequality was obtained for $m > 4/\varepsilon$ by applying (7.2.7) a total of $[m\varepsilon/2]$ times. The result follows immediately since $b(r,m,\gamma) \le 1$.

□

For further discussion of the law of large numbers, see Feller [36], pp. 139-142 and 181, Exercise 18.

Definition 7.2.2. For $\sigma \subseteq \{1,2,\ldots,m\}$, set

$$M(\sigma,\varepsilon) = M_m(\sigma,\varepsilon) = \inf \sum_{i\in\sigma} x_i,$$

where the infimum is over all $\underline{x} = (x_1,\ldots,x_m) \in M_\varepsilon(m)$.

If a point is ε-random, then certainly (ε small, m large) the sum of its coordinates cannot be too small. This is made precise by the following lemma:

Lemma 7.2.1. Let $\sigma_1,\ldots,\sigma_J \subseteq \{1,2,\ldots,m\}$. If $Jm/n \le \Sigma_{j=1}^{J} |\sigma_j|$, then

$$\frac{(1 - 2\varepsilon n)Jm}{2n^2} \le \sum_{j=1}^{J} M(\sigma_j,\varepsilon).$$

Proof. Say that $\underline{x} \in M_\varepsilon(m)$ and that its coordinates are $x_{i_1} \le x_{i_2} \le \cdots \le x_{i_m}$. On replacing the γ of Definition 7.2.1 with x_{i_t}, it follows that

$$m(x_{i_t} - \varepsilon) < t < m(x_{i_t} + \varepsilon),$$

so $t/m - \varepsilon < x_{i_t} < t/m + \varepsilon$. Hence

$$M(\sigma_j,\varepsilon) \ge \sum_{t=1}^{|\sigma_j|} \left(\frac{t}{m} - \varepsilon\right) = \frac{|\sigma_j|(|\sigma_j| + 1)}{2m} - \varepsilon|\sigma_j|.$$

By the Cauchy-Schwarz inequality,

$$\sum_{j=1}^{J} 1^2 \sum_{j=1}^{J} |\sigma_j|^2 \ge \frac{J^2m^2}{n^2},$$

so

$$\sum_{j=1}^{J} M(\sigma_j,\epsilon) \geq \sum_{j=1}^{J} \frac{|\sigma_j|^2}{2m} + \frac{Jm}{2mn} - \frac{\epsilon Jm}{n}$$

$$\geq \frac{Jm^2}{2mn^2} - \frac{\epsilon Jm}{n} = \frac{(1 - 2\epsilon n)Jm}{2n^2} .$$

□

7.3 A Geometric Representation of m-Variable Polynomials

In the following it will be useful to think of the polynomials of degree at most r_i in x_i $(i = 1,\ldots,m)$ as being mapped into the point set $[0,1]^m$ as follows:

$$P = \sum_{i_1=0}^{r_1} \cdots \sum_{i_m=0}^{r_m} a_{i_1 \cdots i_m} x_1^{i_1} \cdots x_m^{i_m} \rightarrow \{(\frac{i_1}{r_1},\ldots,\frac{i_m}{r_m}) \mid a_{i_1 \cdots i_m} \neq 0\}.$$

The next lemma tells us that we can find a certain "not too large" polynomial $P \in \underset{\sim}{Z}[x_1,\ldots,x_m]$ such that $P(x_1 + \alpha,\ldots,x_m + \alpha)$ is mapped into ϵ-random points only. Here $C_i = C_i(\alpha)$, $i = 1,2,\ldots,$ is used to denote a positive constant that depends on α alone.

Lemma 7.3.1. Let α be an algebraic integer and $\epsilon > 0$. Then there is an $m_0 = m_0(\epsilon,\alpha)$ and an $r = r(m,\epsilon)$ such that if $m_0 \leq m$ and $r \leq r_i$ for $i = 1,\ldots,m$, then there exists some $P \in \underset{\sim}{Z}[x_1,\ldots,x_m]$ with $P \not\equiv 0$ such that

(i) $d_{x_i}(P) \leq r_i$, $i = 1,\ldots,m$.

(ii) $H(P) \leq C_1^{r_1+\cdots+r_m}$.

(iii) $P(x_1 + \alpha,\ldots,x_m + \alpha) = \Sigma_{i_1=0}^{r_1}\cdots\Sigma_{i_m=0}^{r_m} a_{i_1\cdots i_m} x_1^{i_1}\cdots x_m^{i_m}$,

where

(a) $|a_{i_1\cdots i_m}| \leq C_2^{r_1+\cdots+r_m}$.

(b) $a_{i_1\cdots i_m} = 0$ if $(i_1/r_1,\ldots,i_m/r_m) \notin M_\varepsilon(m)$.

<u>Proof</u>. Write

$$(7.3.1)\qquad P(x_1 + \alpha,\ldots,x_m + \alpha) = \sum^{r_1}\cdots\sum^{r_m} b_{j_1\cdots j_m}(x_1 + \alpha)^{j_1}\cdots (x_m + \alpha)^{j_m},$$

where the $b_{j_1\cdots j_m}$ are undetermined integers. Relabel these integers as $b_1,\ldots,b_N$, where $N = (r_1 + 1)\cdots(r_m + 1)$. Then

$$a_{i_1\cdots i_m} = \sum_{i=0}^{r_1+\cdots+r_m} \sum_{j=1}^{N} c_{ij}b_j\alpha^i,$$

where $|c_{ij}| \leq 2^{r_1+\cdots+r_m}$. Now, by Lemma 4.1.4, we have

$$\alpha^k = \sum_{i=0}^{d(\alpha)-1} e_{ki}\alpha^i, \qquad |e_{ki}| \leq (H + 1)^k,$$

provided $d(\alpha) \leq k$ and $H = H(\alpha)$. Here $c_{ij}, e_{ki} \in \underset{\sim}{Z}$. Thus

$$a_{i_1\cdots i_m} = \sum_{i=0}^{d(\alpha)-1} \left(\sum_{j=1}^{N} g_{ij}b_j\right)\alpha^i,$$

where

$$|g_{ij}| \le 2^{r_1+\cdots+r_m}[1 + (r_1 + \cdots + r_m)(H + 1)^{r_1+\cdots+r_m}]$$
$$\le C_3^{r_1+\cdots+r_m},$$

so to make a given $a_{i_1\cdots i_m}$ vanish we have to make $d(\alpha)$ linear forms in $b_1,\ldots,b_N$ with integer coefficients vanish. Fix a $\delta > 0$. Then by Theorem 7.2.1 there is an m_0 such that, for $m \ge m_0$,

$$\mu[M_\epsilon(m)] \ge 1 - \delta.$$

For such an m let $T_r(m) = T_r(m;r_1,\ldots,r_m)$ denote the number of m-tuples $(i_1,\ldots,i_m)$ such that

$$(i_1/r_1,\ldots,i_m/r_m) \notin M_\epsilon(m),$$

where i_j $(j = 1,\ldots,m)$ is an integer satisfying $0 \le i_j \le r_j$. Since $M_\epsilon(m)$ consists solely of m-dimensional parallelopipeds, it is clear (e.g., from the definition of the m-dimensional Riemann integral) that

$$\frac{T_r(m)}{(r_1 + 1)\cdots(r_m + 1)} \to \mu\{[0,1]^m - M_\epsilon(m)\} \le \delta$$

as $r \to \infty$. Hence there exists an $r = r(m,\epsilon)$ such that $r_i \ge r$, $i = 1,\ldots,m$, implies that

$$T_r(m) \le 2\delta N.$$

So to make all the corresponding $a_{i_1\cdots i_m}$ vanish it suffices to make

$$M = 2\delta N d(\alpha)$$

linear forms in $b_1,\ldots,b_N$ vanish. By an appropriate choice of δ [e.g., $\delta < 1/4\, d(\alpha)$] we can make $2M < N$, so the pigeonhole principle (Lemma 2.2.1) applies. Thus there are $b_1,\ldots,b_N \in \underset{\sim}{Z}$ not all zero such that these forms vanish, and

$$|b_i| \le (NC_3^{r_1+\cdots+r_m})^{M/(N-M)} \le (2C_3)^{r_1+\cdots+r_m}.$$

The result now follows, since part (ii) of the lemma holds with $C_1 = 2C_3$ and by (7.3.2)

$$|a_{i_1\cdots i_m}| \le d(\alpha)NC_3^{r_1+\cdots+r_m}C_1^{r_1+\cdots+r_m}C_4(\alpha)$$
$$\le C_2^{r_1+\cdots r_m}.$$

□

7.4. If α Were Easy To Approximate, Then P Would Have a Large Index

It is easy to show the following result:

Lemma 7.4.1. If $(x_1,\ldots,x_m) \in M_\epsilon(m)$ and $|\eta_i| \le \epsilon$ for $1 \le i \le m$, then $(x_1 + \eta_1,\ldots,x_m + \eta_m) \in M_{2\epsilon}(m)$.

In the following $C_i = C_i(\alpha,\lambda,n)$ denotes a positive constant that depends only on α, λ, and the integer n.

Theorem 7.4.1 (Wirsing). Let $\lambda > 2n$ and let ϵ be so small that

$$1 + \epsilon - \frac{\lambda}{2n}(1 - 4\epsilon n) < -\epsilon. \tag{7.4.1}$$

Assume that $m_0(\epsilon,\alpha) \le m$ and that the integers $r_1,\ldots,r_m$ satisfy

$r_m \leq \cdots \leq r_2 \leq r_1$ and $r(m,\epsilon) < r_i$ for all i. Let $\beta_1,\ldots,\beta_m$ be algebraic numbers of degree n satisfying

(i) $|\alpha - \beta_i| < 1/H(\beta_i)^\lambda$, $i = 1,\ldots,m$.

(ii) $C_{11} < H(\beta_1)^\epsilon$.

(iii) $r_1 \log H(\beta_1) \leq r_i \log H(\beta_i) \leq (1 + \epsilon) r_1 \log H(\beta_1)$, $i = 1,\ldots,m$.

Then

$$\epsilon < I(P; r_1,\ldots,r_m; \beta_1,\ldots,\beta_m), \tag{7.4.2}$$

where P is any polynomial produced by Lemma 7.3.1.

Note: The constant C_{11} will be determined in the course of the proof. Observe that

$$m(\tfrac{1}{2} - \epsilon) \leq I(P) \leq m(\tfrac{1}{2} + \epsilon)$$

since P maps into an ϵ-random point. Theorem 7.4.1 says that the index of P is also bounded away from zero at the point $(\beta_1,\ldots,\beta_m)$, which is very near to $(\alpha,\ldots,\alpha)$. This is nontrivial since the index is not a continuous function. Inequality (7.4.2) will ultimately contradict Roth's lemma if (i) has infinitely many solutions.

Proof. By Lemma 7.3.1(iii),

$$P_{j_1\cdots j_m}(x_1 + \alpha,\ldots,x_m + \alpha) = \Sigma\cdots\Sigma\, a_{i_1\cdots i_m} \binom{i_1}{j_1}\cdots\binom{i_m}{j_m} \cdot x_1^{i_1 - j_1}\cdots x_m^{i_m - j_m},$$

where

$$(7.4.3) \qquad P_{j_1\cdots j_m}(x_1,\ldots,x_m) = \left(\frac{\partial^{j_1+\cdots+j_m}}{j_1!\cdots j_m!\,\partial x_1^{j_1}\cdots\partial x_m^{j_m}}\right)P$$

$$\in \underset{\sim}{Z}[x_1,\ldots,x_m],$$

so

$$(7.4.4) \qquad P_{j_1\cdots j_m}(\beta_1,\ldots,\beta_m) = \sum_{i_1}\cdots\sum_{i_m} a_{i_1\cdots i_m}\binom{i_1}{j_1}\cdots\binom{i_m}{j_m}$$

$$\cdot\,(\beta_1-\alpha)^{i_1-j_1}\cdots(\beta_m-\alpha)^{i_m-j_m}.$$

Now assume that $j_1/r_1 + \cdots + j_m/r_m \leq \epsilon$; it then suffices to show that

$$P(\beta) = P_{j_1\cdots j_m}(\beta_1,\ldots,\beta_m) = 0.$$

Whenever

$$(7.4.5) \qquad \left(\frac{i_1 - j_1}{r_1},\ldots,\frac{i_m - j_m}{r_m}\right) \in M_{2\epsilon}(m)$$

is not valid, $(i_1/r_1,\ldots,i_m/r_m) \notin M_\epsilon(m)$ by Lemma 7.4.1, so $a_{i_1\cdots i_m} = 0$ by condition (b) of Lemma 7.3.1(iii). Hence we may assume that (7.4.5) holds for every nonzero term on the right-hand side of (7.4.4).

Now $P(\beta) \in \underset{\sim}{K}$, where $\underset{\sim}{K} = \underset{\sim}{Q}(\beta_1,\ldots,\beta_m)$. Set $D = [\underset{\sim}{K}:\underset{\sim}{Q}]$; clearly $n\,|\,D$ and $n \leq D \leq n^m$. Let $B_1,\ldots,B_m$ denote the lead coefficients of $\beta_1,\ldots,\beta_m$ and $\tau_1,\ldots,\tau_D$ the isomorphisms of $\underset{\sim}{K}$ into $\underset{\sim}{C}$ that fix $\underset{\sim}{Q}$. Then by (7.4.3)

$$\tau_k[P_{j_1 \cdots j_m}(\beta_1, \ldots, \beta_m)] = P_{j_1 \cdots j_m}(\tau_k(\beta_1), \ldots, \tau_k(\beta_m))$$

for $1 \leq k \leq D$. By the symmetric function theorem,

$$\Pi_P = \prod_{k=1}^{D} \tau_k[P_{j_1 \cdots j_m}(\beta_1, \ldots, \beta_m)] \in \underset{\sim}{Q},$$

and by Lemma 4.1.1

$$z \equiv B_1^{r_1 D/n} \cdots B_m^{r_m D/n} \Pi_P \in \underset{\sim}{Z}. \tag{7.4.6}$$

We shall show that $|z| < 1$, hence $z = 0$, $\Pi_P = 0$, and the result follows. Now

$$|\tau_k[P_{j_1 \cdots j_m}(\beta_1, \ldots, \beta_m)]| \leq C_5^{r_1 + \cdots + r_m} \max_{\underset{\sim}{y} \in M} \tag{7.4.7}$$

$$|\tau_k(\beta_1) - \alpha|^{y_1 r_1} \cdots |\tau_k(\beta_m) - \alpha|^{y_m r_m},$$

where $M = M_{2\epsilon}(m)$. If $\tau_k(\beta_i) = \beta_i$, we have

$$|\tau_k(\beta_i) - \alpha|^{y_i r_i} \leq H(\beta_i)^{-r_i \lambda y_i} \leq H(\beta_1)^{-r_1 \lambda y_i}$$

by (i) and (iii) while the crude estimate

$$|\tau_k(\beta_i) - \alpha|^{y_i r_i} \leq [C_6 + |\tau_k(\beta_i)|]^{r_i}$$

is always valid for some $C_6 > 1$. Thus

$$|\Pi_P| \leq C_5^{D(r_1 + \cdots + r_m)} \prod_{k=1}^{D} \{ \prod_{i=1}^{m} [C_6 + |\tau_k(\beta_i)|]^{r_i} \cdot \max_{\underset{\sim}{y} \in M} \prod_{i \in \sigma_k} H(\beta_1)^{-r_1 \lambda y_i} \},$$

where $\sigma_k = \{i \mid 1 \le i \le m,\ \tau_k(\beta_i) = \beta_i\}$. By elementary Galois theory,

$$\sum_{k=1}^{D} |\sigma_k| = \frac{Dm}{n}$$

(see Fig. 5). By (7.4.6) we have

$$|z| \le C_5^{D(r_1+\cdots+r_m)} \cdot \prod_{i=1}^{m} \{B_i^{D/n} \prod_{k=1}^{D} [C_6 + |\tau_k(\beta_i)|]\}^{r_i}$$

$$\cdot \prod_{k=1}^{D} \max_{\underline{y}\in M} \prod_{i\in\sigma_k} H(\beta_1)^{-r_1\lambda y_i} \equiv A_1 \cdot A_2 \cdot A_3.$$

$$\begin{matrix} \tau_1(\beta_1) & \cdots & \tau_1(\beta_m) \\ \vdots & & \vdots \\ \tau_D(\beta_1) & \cdots & \tau_D(\beta_m) \end{matrix}$$

Figure 5

Now from Definition 7.2.2 and Lemma 7.2.1 (with J replaced by D),

$$\log A_3 \le -r_1\lambda \log H(\beta_1) \sum_{k=1}^{D} M(\sigma_k, 2\epsilon)$$

$$\le -r_1\lambda \log H(\beta_1) \frac{DM}{2n^2} (1 - 4\epsilon n),$$

and from Theorem 4.1.2(i) and hypothesis (iii), we have

$$A_2 \le \prod_{i=1}^{m} \{B_i^{D/n} 2^D C_7^D \prod_{k=1}^{D} |\tau_k(\beta_i)|\}^{r_i}$$

$$\le \prod_{i=1}^{m} \{2^D C_7^D B_i^{D/n} (\frac{L(\beta_i)}{B_i})^{D/n}\}^{r_i}$$

$$\le C_8^{mDr_1} \prod_{i=1}^{m} \{H(\beta_1)^{r_1(1+\varepsilon)D/n}\}$$

$$= C_8^{mDr_1} H(\beta_1)^{mDr_1(1+\varepsilon)/n}$$

since $r_i \le r_1$, $1 \le i \le m$, and

$L(\beta_i)^{r_i} \le C_9^{r_i} H(\beta_i)^{r_i} \le C_9^{r_i} H(\beta_1)^{r_1(1+\varepsilon)}$. Hence

$$|z| \le C_{10}^{mDr_1} H(\beta_1)^{(mDr_1/n)\{1+\varepsilon-(\lambda/2n)(1-4\varepsilon n)\}}$$

$$< \{C_{10}^n / H(\beta_1)^{\varepsilon}\}^{Dmr_1/n}$$

by (7.4.1). The fact that $|z| < 1$ follows by taking $C_{11} = 2C_{10}^n$.

□

The results of Theorems 6.6.2 and 7.4.1 go in opposite directions. The idea now is to assume that there are infinitely many "good" approximations to α, namely, $\beta_1,\ldots,\beta_m,\ldots$, and to choose from these a subsequence satisfying the hypotheses of both theorems. The resulting contradiction will show that there cannot be infinitely many such β_i.

7.5. Proof of Wirsing's Theorem

Just as in the proof of Thue's theorem, it suffices to show that

$$|\alpha - \beta_i| < 1/H(\beta_i)^{\lambda} \qquad (2n < \lambda) \tag{7.5.1}$$

has only finitely many solutions β_i with $d(\beta_i) \le n$ when α is an algebraic integer. Theorems 6.6.2 and 7.4.1 are restated here as Theorems 7.5.1 and 7.5.2.

Theorem 7.5.1. Let $P \in \underset{\sim}{Z}[x_1,\ldots,x_m]$, $P \not\equiv 0$, $H(P) < C^{r_1 m}$, and $d_{x_i}(P) \le r_i$. If $0 < t \le 1$ and

(i) $r_{j+1} < tr_j$, $1 \le j \le m - 1$,

(ii) $r_1 \log H(\beta_1) \le r_j \log H(\beta_j)$, $1 \le j \le m$,

(iii) $2^{m(m+1)(2m+3)/t} < H(\beta_1)$,

(iv) $C^{r_1 m} \le H(\beta_1)^{r_1 t/(m+1)}$,

then

$$I(P;r_1,\ldots,r_m;\beta_1,\ldots,\beta_m) \le 2^{m+1}t^{2^{-m+1}}. \tag{I}$$

Theorem 7.5.2. Let the β_i satisfy (7.5.1). If

(i) $1 + \epsilon - (\lambda/2n)(1 - 4\epsilon n) < -\epsilon$,

(ii) $m_0(\epsilon,\alpha) < m$,

(iii) $r_m \le \cdots \le r_1$,

(iv) $r(m,\epsilon) < r_i$, $1 \le i \le m$,

(v) $C_{11} < H(\beta_1)^{\epsilon}$,

(vi) $r_1 \log H(\beta_1) \le r_i \log H(\beta_i) \le (1 + \epsilon)r_1 \log H(\beta_1)$, $1 \le i \le m$, then there exists a polynomial P as in Theorem 7.5.1 such that

(II) $$\varepsilon < I(P; r_1, \ldots, r_m; \beta_1, \ldots, \beta_m).$$

First choose ε satisfying condition (i) of Theorem 5.7.2, then an $m > m_0(\varepsilon, \alpha)$ for condition (ii) and then a value of t so small that

(7.5.2) $$2^{m-1} t^{2^{-m+1}} < \varepsilon.$$

Choose β_1 to satisfy conditions (iii) and (iv) of Theorem 7.5.1 and condition (v) of Theorem 7.5.2. Consider the intervals

$$[a_i, b_i] = \left[\frac{\log H(\beta_1)}{\log H(\beta_i)}, \ (1 + \varepsilon)\frac{\log H(\beta_1)}{\log H(\beta_i)} \right].$$

Choose each successive β_i so that these do not overlap, and in fact so that $b_{i+1} < ta_i$. Choose r_1 so large that

$$r(m, \varepsilon) < \frac{r_1 \log H(\beta_1)}{\log H(\beta_m)}$$

and $[r_1 a_i, r_1 b_i]$ contains an integer r_i for $i = 2, \ldots, m$. Then conditions (i) and (ii) of Theorem 7.5.1 and conditions (iv) [since $r(m, \varepsilon) < r_m$] and (vi) of Theorem 7.5.2 follow. For this choice of parameters (I) and (II) both hold, so (7.5.2) is a contradiction.

□

<u>Note</u>: Most of this proof is based on an unpublished manuscript of W. Schmidt's on Wirsing's theorem. Where this manuscript quotes earlier work we have followed a variety of sources: Cassels [18],

pp. 103-119, LeVeque [58], pp. 121-152, Mahler [67], pp. 77-106, and the original paper by Roth [93].

Exercises

7.5.1. Let $b \in \underset{\sim}{Z}^+$ be fixed. Show that if $\alpha \in \underset{\sim}{A}$ and $2 \le d(\alpha)$, then

$$|\alpha - \frac{p}{q}| < q^{-\mu}$$

has only finitely many solutions p_k/q_k with $p_k, q_k \in \underset{\sim}{Z}$ and $q_k = q_k' b^{s_k}$, where s_k is a nonnegative integer, provided that

$$\lim\sup(\log q_k'/\log q_k) = \eta < \mu - 1.$$

(Hint (Schneider [103], pp. 13-34): It clearly suffices to show this when $(q_k', b) = 1$ for all k. The idea is to assume that there is an infinite sequence of solutions $\{p_k/q_k\}$ and to choose from it m elements (relabel them $p_1/q_1, \ldots, p_m/q_m$, where $q_1 < q_2 < \ldots < q_m$) such that (i) $P_{j_1 \cdots j_m}(p_1/q_1, \ldots, p_m/q_m) \neq 0$ follows from Roth's lemma, where P is a polynomial produced by Lemma 7.3.1, and (ii) the obvious upper and lower bounds for $|P_{j_1 \cdots j_m}(p_1/q_1, \ldots, p_m/q_m)| = |P_j|$ contradict one another. (This is the technique described in Section 6.2.) Clearly, for the lower bound,

$$q_1^{-r_1} \cdots q_m^{-r_m} = q_1'^{-r_1} \cdots q_m'^{-r_m} b^{-r_1 s_1 - \cdots - r_m s_m} \le |P_j|,$$

but this can be improved to

$$\beta q_1^{-r_1} \cdots q_m^{-r_m} \le |P_j|$$

since the power of b occurring in the denominator cannot exceed

$$b^s = \max b^{i_1 s_1 + \cdots + i_m s_m}$$

$$= b^{r_1 s_1 + \cdots + r_m s_m}[\min b^{(r_1 - i_1)s_1 + \cdots + (r_m - i_m)s_m}]^{-1}$$

$$= b^{r_1 s_1 + \cdots + r_m s_m}\beta^{-1},$$

where the maximum and minimum are over all $(i_1, \ldots, i_m)$ such that $\Sigma i_j/r_j \leq m(1/2 + \epsilon)$. Thus for $\epsilon > 0$ and q_1 sufficiently large, it follows from the definition of limit superior that

$$\beta \geq \min[q_1^{(r_1 - i_1)} \cdots q_m^{(r_m - i_m)}]^{(1-\eta-\epsilon)}$$

$$\geq \min[q_1^{(r_1 - i_1)/r_1 + \cdots + (r_m - i_m)/r_m}]^{(1-\eta-\epsilon)r_1}$$

if condition (iii) of Theorem 7.4.1 holds. Thus

$$\beta \geq \min\{q_1^{m - m(1/2+\epsilon)}\}^{(1-\eta-\epsilon)r_1} = q_1^{r_1 m(1/2-\epsilon)(1-\eta-\epsilon)}$$

and

$$q_1^{-r_1 m(1/2+5\epsilon/2+\eta/2)} \leq q_1^{-r_1 m(1+\epsilon)+r_1 m(1/2-\epsilon)(1-\eta-\epsilon)} \leq |P_j|.$$

For the upper bound, use (7.4.4) with $\beta_1 = p_1/q_1, \ldots, \beta_m = p_m/q_m$. If condition (iii) of Theorem 7.4.1 holds, it follows from Lemma 7.3.1 and our hypothesis that

$$|P_j| \le C^{r_1+\cdots+r_m} q_1^{-\mu(i_1-j_1)} \cdots q_m^{-\mu(i_m-j_m)}$$

$$\le C^{r_1+\cdots+r_m} q_1^{-\mu r_1\{(i_1-j_1)/r_1+\cdots+(i_m-j_m)/r_m\}}$$

$$\le C^{r_1+\cdots+r_m} q_1^{-\mu r_1(i_1/r_1+\cdots+i_m/r_m-J)}$$

$$\le C^{r_1+\cdots+r_m} q_1^{-\mu r_1 m(1/2-\epsilon)+\mu r_1 J},$$

where $J = \Sigma j_i/r_i \le 2^{m+1} t^{2^{-m+1}}$ is made negligibly small by an appropriate choice of t. By combining the upper and lower bounds, we find that

$$\mu(1 - 2\epsilon) \le \frac{2 \log C}{\log q_1} + 1 + \eta + 5\epsilon + \frac{2\mu J}{m},$$

but for ϵ and J sufficiently small and q_1 sufficiently large, this is a contradiction. Note that if b = 1, the result of this exercise is simply Roth's theorem.)

7.5.2. Show that $\Sigma_{n=0}^{\infty} 2^{-2^n}$ is transcendental. This "improves" Exercise 3.2.20. (Hint: Use Exercise 7.5.1.)

7.5.3. Let S be the set of all tuples $(i_1,\ldots,i_m)$ with $0 \le i_j \le r_j$ such that

$$\epsilon m \le \left| \sum_{h=1}^{m} \frac{i_h}{r_h} - \frac{m}{2} \right| = \left|\Sigma - \frac{m}{2}\right| \qquad (0 < \epsilon < 4).$$

Show that $|S| \le 2(r_1 + 1)\cdots(r_m + 1)e^{-\epsilon^2 m/4}$. (Hint: Let T be the set of tuples in S for which $\Sigma - m/2$ is positive. Then

$$|T| \exp(\frac{\varepsilon^2 m}{2}) \le \sum_T \exp\{\frac{\varepsilon}{2} (\sum_{j=1}^{m} \frac{i_j}{r_j} - \frac{m}{2})\}$$

$$\le \prod_{j=1}^{m} (\sum_{i=0}^{r_j} \exp\{\frac{\varepsilon}{2} (\frac{i}{r_j} - \frac{1}{2})\}) = \prod_{j=1}^{m} (\Sigma)_j ,$$

and since $e^x \le 1 + x + x^2$ for $x \in \underset{\sim}{R}$ and $|x| \le 1$, we have

$$\Sigma_j \le \sum_{i=0}^{r_j} \{1 + \frac{\varepsilon}{2} (\frac{i}{r_j} - \frac{1}{2}) + \frac{\varepsilon^2}{4}\} \le (r_j + 1)\ (1 + \frac{\varepsilon^2}{4})$$

$$\le (r_j + 1) \exp(\frac{\varepsilon^2}{4}).$$

This is a specialization of Schmidt's ([100], pp. 65-70) generalization of Mahler's ([67], pp. 163-168) exposition of G. Reuter's improvement of Schneider's [102] result.)

7.5.4. Let S be the set of all tuples $(i_1,\ldots,i_m)$ with $0 \le i_j \le r_j$ such that

$$\varepsilon \sum_{h=1}^{m} \frac{r_h}{\rho_h} \le |\sum_{h=1}^{m} \frac{i_h}{\rho_h} - \frac{1}{2} \Sigma \frac{r_h}{\rho_h}|,$$

where $0 < \varepsilon < 2$ and $|r_h/\rho_h - 1| < 1/10$. Show that

$$|S| \le 2(r_1 + 1)\cdots(r_m + 1)e^{-\varepsilon^2 m/5}.$$

(Hint: Proceed as in Exercise 7.5.3.)

7.5.5. Let α be a real algebraic number. Show that there is a constant $C = C(\alpha) > 0$ such that if $(p_i,q_i) = 1$ for $i \ge 1$ and p_i/q_i is an infinite sequence of rational solutions of

$$|\alpha - \frac{p}{q}| < q^{-2-C(\alpha)(\log\log\log q)^{-1/2}},$$

then

$$\limsup \frac{\log q_{i+1}}{\log q_i} = \infty.$$

(This result of Cugiani's has been generalized by Mahler [67], pp. 169-180, who also proves it with $C(\alpha) = 9d(\alpha)$.) (Hint: As in Exercise 7.5.1, the idea is to get conflicting upper and lower bounds on some nonvanishing derivative of the polynomial P of (7.4.4). (In the notation of Theorem 7.4.1, n = 1 and $\beta_j = p_j/q_j$.) However, instead of using the law of large numbers in the construction of P, it is expedient to use Exercise 7.5.4; thus for fixed $\rho_1,\ldots,\rho_m$ with

$$|r_h/\rho_h - 1| < 1/10,$$

we have

$$a_{i_1\cdots i_m} = 0$$

unless

(*)
$$\left|\Sigma \frac{i_h}{\rho_h} - \frac{1}{2} \Sigma \frac{r_h}{\rho_h}\right| \leq \epsilon \sum_{h=1}^{m} \frac{r_h}{\rho_h}.$$

If Roth's lemma is applicable, there will be $j_1,\ldots,j_m$ such that $P_{j_1\cdots j_m}(p_1/q_1,\ldots,p_m/q_m) \neq 0$ and

$$J = \Sigma \frac{j_i}{r_i} \leq 2^{m+1} t^{2^{-m+1}}.$$

Take $t = \exp(-m2^{m-1})$. Now if the conclusion of this exercise is false, it is possible (for m sufficiently large) to pick a finite subsequence $p_1/q_1,\ldots,p_m/q_m$ out of the original sequence $\{p_i/q_i\}$

such that

$$\exp\left(\frac{2m^3}{t}\right) \le H_h = H\left(\frac{p_h}{q_h}\right) \le \exp\exp\exp m$$

and

$$\frac{\log H_{h+1}}{\log H_h} \ge 2/t \qquad (h = 1,\dots,m).$$

One reason we want the rather large lower bound on every H_h is so that the hypotheses of Roth's lemma will be fulfilled. Next choose r_1 and then $r_2,\dots,r_m$ so that

$$(r_h - 1)\log H_h \le r_1 \log H_1 \le r_h \log H_h \qquad (h = 1,\dots,m)$$

and $\theta = \max(r_h - 1)^{-1}$ satisfies $0 < \theta \le 1/m$. Then $r_h < tr_{h-1}$ and $r_h \log H_h \le (1 + \theta) r_1 \log H_1$. Finally, let

$$\rho_h = 2r_h/\{2 + \epsilon(H_h)\},$$

where $\epsilon(H) = C(\alpha)(\log\log\log H)^{-1/2}$. We now have the lower bound

$$\begin{aligned} |P_j| &= |P_{j_1\cdots j_m}(p_1/q_1,\dots,p_m/q_m)| \\ &\ge (H_1^{r_1-j_1}\cdots H_m^{r_m-j_m})^{-1} = (H_1^{(r_1-j_1)/r_1}\cdots H_m^{(r_m-j_m)/r_1})^{-r_1} \\ &\ge H_1^{-r_1(1+\theta)(m-\Sigma\, j_h/r_h)} \end{aligned}$$

and the upper bound

$$|P_j| \le C_2^{mr_1}(\alpha) \max \prod_{h=1}^{m} H_h^{-(i_h-j_h)\{2+\epsilon(H_h)\}}$$

$$\le C_2^{mr_1} \max \Pi H_h^{-(i_h-j_h)2r_h/\rho_h}$$

$$\le C_2^{mr_1} \max H_1^{-2r_1\Sigma_{h=1}^{m}(i_h-j_h)/\rho_h}$$

$$\le C_2^{mr_1} H_1^{-2r_1\{(1/2-\epsilon)\Sigma r_h/\rho_h - \Sigma j_h/\rho_h\}},$$

where the maximum was taken over all $(i_1,\ldots,i_m)$ satisfying (*). For the construction of P in Lemma 7.3.1 to be valid we need $2\exp(-\epsilon^2 m/5) < 1$. On the other hand, the upper bound just obtained for $|P_j|$ is strongest when ϵ is smallest. Hence we choose $\epsilon = C_3 m^{-1/2}$, where $C_3 = C_3(\alpha)$ is large enough.

Now θ and $\Sigma\, j_h/\rho_h$ are negligible, and

$$\Sigma \frac{r_h}{\rho_h} = \sum_{h=1}^{m} \left(1 + \frac{\epsilon(H_h)}{2}\right) \ge m + C_4(\alpha)m^{1/2},$$

so by combining the upper and lower bounds we essentially obtain

$$H_1^{C_4-2C_3} \le C_2^{\sqrt{m}} .$$

By making the original $C(\alpha)$ sufficiently large we can ensure that $C_4 > 2C_3$, and then the above inequality will contradict the original choice of H_1 for m sufficiently large.)

**<u>7.5.6</u>. (A. S. Fraenkel) Let $P = P(x_1,\ldots,x_m) \in \underset{\sim}{Z}[x_1,\ldots,x_m]$ be of degree r_i in x_i. Does it follow that

$$I(P;r_1,\ldots,r_m;\alpha,\ldots,\alpha) \leq m/2$$

for any irrational algebraic α?

**7.5.7. Let α be a real algebraic number, $3 \leq d(\alpha)$. Does there exist a constant $C = C(\alpha) > 0$ (perhaps very small) such that

$$|\alpha - \frac{p}{q}| > C/q^2?$$

(Hint: If C exists, this problem must be very difficult; the best hope is that one can disprove its existence.)

7.5.8. Show that the $C > 0$ of Exercise 7.5.7 exists if and only if there is some $C_1(\alpha) > 0$ (perhaps very large) such that

$$\alpha = a_0 + \frac{1}{a_1} + \frac{1}{a_2} + \ldots,$$

where $a_i \in \underset{\sim}{Z}^+$ and $|a_i| < C_1(\alpha)$ for all $i \in \underset{\sim}{Z}^+$. (Note: If α is the real root of $x^3 - 8x - 10$, then $a_{16} = 3$ and $a_{32} = 1$, but $a_{17} = 22{,}986$ and $a_{33} = 1{,}501{,}790$. See Stark [119].)

*7.5.9. Let $\psi(x)$ be a positive, nonincreasing function defined for $x \geq 1$. Show that

$$|\alpha - \frac{p}{q}| < \psi(q)/q$$

has finitely or infinitely many solutions $p,q \in \underset{\sim}{Z}$, $q \geq 1$, for almost all real α according to whether

$$\sum_{n=1}^{\infty} \psi(n)$$

converges or diverges. (Hint: A proof of this theorem of Khinchin's can be found in Cassels [18], pp. 120-132.)

Since $\Sigma_{q=1}^{\infty} q^{-1-\epsilon}$ converges, Roth's theorem "merely" asserts that real algebraic irrationalities behave in the same manner as "typical" real numbers. W. Schmidt has commented that this sort of reasoning, together with the result of Exercise 7.5.9, would lead us to expect that (i)

$$|\alpha - \frac{p}{q}| < 1/q^2(\ell n\ q)^{1+\epsilon},$$

where α is a real algebraic irrationality, has only a finite number of solutions for any $\epsilon > 0$, whereas (ii)

$$|\alpha - \frac{p}{q}| < \delta/q^2$$

(or even $|\alpha - p/q| < 1/q^2\ \ell n\ q$) should have infinitely many, no matter how small δ is. Of course, we must exclude quadratic irrationalities in (ii). At this date both (i) and (ii) seem unapproachable; some numerical evidence in favor of (ii) can be found in Bryuno [15], von Neumann and Tuckerman [77], and Richtmyer et al. [88]. See also Lang and Trotter [56].

Chapter 8

MOTIVATION FOR THE PROOFS OF VARIOUS INEQUALITIES INVOLVING ALGEBRAIC AND TRANSCENDENTAL NUMBERS

8.1. Generalities

Many proofs that a given number α cannot be algebraic of degree n fall into a pattern roughly similar to the following: Assume that $d(\alpha) = n$. Then $f(k) = k|S(\alpha_1,\ldots,\alpha_n)| \in \underset{\sim}{Z}^+$ for $k \in \underset{\sim}{Z}^+$ sufficiently large, where $S = S(\alpha_1,\ldots,\alpha_n)$ is a symmetric function of $\alpha_1,\ldots,\alpha_n$, the conjugates of α. But $f(k) < 1$, a contradiction. For example, assume that $d(e) = 1$, where e is the base of natural logarithms. Then

$$f(k) = (2k - 1)!\,|e^{-1} - \sum_{n=0}^{2k-1} \frac{(-1)^n}{n!}| \in \underset{\sim}{Z}^+$$

for $k \in \underset{\sim}{Z}^+$ and k sufficiently large. But

$$0 < f(k) = (2k - 1)! \sum_{n=2k}^{\infty} \frac{(-1)^n}{n!} < \frac{1}{2k} < 1,$$

a contradiction.

Similarly, to obtain an inequality involving any algebraic numbers $\alpha_1,\ldots,\alpha_n$ one could assume the <u>negation</u> of this inequality and attempt to use it to construct an auxiliary function $f(z) = f(z;\alpha_1,\ldots,\alpha_n)$ such that for some z_0 both $f(z_0) \in \underset{\sim}{Z}$ and

$0 < |f(z_0)| < 1$ are valid--a contradiction. A slight variation on this theme is to construct an f such that $f(z) \in \mathbb{Z}$ when $z \in \mathbb{Z}$, $z \geq 0$, and

$$|f(z)| < 1 \quad \text{when} \quad z \in \mathbb{Z} \qquad (0 \leq z \leq N). \tag{8.1.1}$$

If f can have at most N consecutive integral zeros beginning with 0, this is a contradiction. This latter condition holds, for example, when

$$f(z) = \sum_{n=0}^{N} c_n f_n(z), \tag{8.1.2}$$

where the c_n are integers not all zero, $f_n(z) \in \mathbb{Z}$ when $z \in \mathbb{Z}$, $z \geq 0$, and

$$D = \det[f_n(z)]_{n,z=0}^{N} \neq 0. \tag{8.1.3}$$

8.2. Flat Functions

Assume now that f(z) is of the form (8.1.2) and that the $f_n(z)$ are analytic (or even entire of exponential type; see comment before Exercise 8.2.5). We shall try to obtain (8.1.1) by making

$$f^{(i)}(0) = \sum_{n=0}^{N} c_n f_n^{(i)}(0)$$

very small for $0 \leq i \leq M - 1$; this can be done for integers $c_n \in Z$ not all zero by some form of the pigeonhole principle. In other words, we are hoping that if some linear combination of $f_1(z),\ldots,f_N(z)$ is very "flat" at $z = 0$, then it remains small in a fairly large real interval $[0,z]$.

In general, the flatness of $f(z)$ at $z = 0$ does not lead to the desired result; for example, if $c_0 = 1$ and $f_0(z) = z^M$ while $c_n = 0$ and $f_n(z) \equiv 0$ for $n \geq 1$, then $f^{(i)}(0) = 0$ for $0 \leq i \leq M - 1$, but (8.1.1) is false for $z \geq 1$. However, Proposition 8.2.1 is an example where this method succeeds to some extent (the $0 \leq z \leq N$ of (8.1.1) is replaced by $0 \leq z \leq [(\ell n\ N)^{m+1+\epsilon}]$).

Definition 8.2.1. Fix $m \geq 0$ and $\epsilon > 0$. Let $\mathfrak{F} = \{f(z) \mid f(z) = \sum_{n=1}^{N} c_n n^z$, where $c_n \in \{-1,0,1\}$, $|f^{(i)}(0)| \leq \exp(-N/16(\ell n\ N)^{m+1+\epsilon})$, and $0 \leq i \leq [(\ell n\ N)^{m+1+\epsilon}]\}$.

Lemma 8.2.1. Let $y_i(\underline{c}) = a_{i1}c_1 + \cdots + a_{iN}c_N$ for $0 \leq i \leq M - 1$ where $\underline{c} = (c_1,\ldots,c_N)$ is an integral vector and $a_{ij} \in \mathbb{R}$, $1 \leq j \leq N$. Say $\sum_{j=1}^{N} |a_{ij}| \leq L$ and $B \in \mathbb{Z}^+$. Then there is a vector $\underline{c} \neq 0$ such that

(i) $|c_i| \leq B$.

(ii) $|y_i(\underline{c})| \leq 2LB/(B + 1)^{N/M} - 2$.

Proof. This is Lemma 2.2.2.

Proposition 8.2.1. $\mathfrak{F} \neq \{0\}$ if N is sufficiently large. Moreover, every $f \in \mathfrak{F}$ satisfies $f(z) = 0$ if $z \in \mathbb{Z}$ and $0 \leq z \leq (\ell n\ N)^m$.

Proof. Set $y_i(\underline{c}) = f^{(i)}(0) = \sum_{n=1}^{N} c_n(\ell n\ n)^i$, $L = N(\ell n\ N)^{M-1}$, $M = 1 + [(\ell n\ N)^{m+1+\epsilon}]$, and $B = 1$. By Lemma 8.2.1 there is a vector $\underline{c} \neq 0$ such that, for $0 \leq i \leq M - 1$, we have

$$|f^{(i)}(0)| \leq \frac{2N(\ell n\ N)^{M-1}}{(2^{N/M} - 2)} \leq \frac{4N(\ell n\ N)^{M-1}}{2^{N/2(M-1)}}$$

$$\leq 4N \exp\{2(\ell n\ N)^{m+1+\epsilon}\ \ell n\ \ell n\ N - (\ell n\ 2)(\frac{N}{4(\ell n\ N)^{m+1+\epsilon}})\}$$

$$\leq \exp\{-(\ell n\ 2)(\frac{N}{8(\ell n\ N)^{m+1+\epsilon}})\}$$

provided N is sufficiently large. Hence $\mathfrak{J} \neq \{0\}$. Next, for $0 \leq z \leq (\ell n\ N)^m$, we have

$$f(z) = \sum_{i=0}^{\infty} \frac{z^i f^{(i)}(0)}{i!} = \sum_{i=0}^{M-2} + \sum_{i=M-1}^{\infty} .$$

Here

$$\left|\sum_{i=0}^{M-2}\right| \leq (M - 1)(\ell n\ N)^{m(M-1)} \exp\{\frac{-N}{16(\ell n\ N)^{m+1+\epsilon}}\}$$

$$\leq \exp\{\frac{-N}{32(\ell n\ N)^{m+1+\epsilon}}\}.$$

In the second sum, $f^{(i)}(0)/i!$ is estimated by means of Cauchy's integral representation:

$$\left|\frac{f^{(i)}(0)}{i!}\right| = \frac{1}{2\pi}\left|\int_{|\zeta|=\rho} f(\zeta)\zeta^{-i-1}\, d\zeta\right| \leq \rho^{-i} N^{\rho+1}.$$

On setting $\rho = (\ell n\ N)^{m+\epsilon}$, we find (again for N sufficiently large) that

$$\left|\sum_{i=M-1}^{\infty}\right| \leq N^{\rho+1} \sum_{i=M-1}^{\infty} (\ln N)^{-i\epsilon}$$

$$\leq 2N \exp\{(\ln N)^{m+1+\epsilon} - \epsilon(M - 1)\ln \ln N\}$$

$$\leq 2N \exp\{-\frac{1}{2}\epsilon(\ln N)^{m+1+\epsilon} \ln \ln N\}.$$

Hence $|f(z)| < 1$ for N sufficiently large and $0 \leq z \leq (\ln N)^m$. But for $z \in \mathbb{Z}$, it is clear that $f(z) \in \mathbb{Z}$, so the result follows.

□

Exercises

8.2.1. Let f be a real-valued differentiable function such that $f(0) = 0$, $f(x) \in \mathbb{Z}$ if $x \in \mathbb{Z}$, and $|f'(x)| < 1$ for $x \geq 0$. Show that $f(x) = 0$ if $x \in \mathbb{Z}$.

8.2.2. Prove a result like Proposition 8.1.1 with $f(z) = z^h \sum_{n=1}^{N} c_n n^z$, where $0 \leq h \leq (\ln N)^{m+1+\epsilon/2}$.

8.2.3. Fix an α with $0 < \alpha < 1$. Let A be a set of N positive integers and say it is false that there exist disjoint sets $A_1, A_2 \subseteq A$ such that

$$\sum_{a_1 \in A} a_1^k = \sum_{a_2 \in A} a_2^k$$

with all $k \in \mathbb{Z}$ with $1 \leq k \leq N^\alpha$. Show then that $\ln m(N) \geq N^{1-2\alpha-\epsilon}$ for N sufficiently large, where m(N) is the largest number in A. (Hint: Let $f(z) = \sum_{n=1}^{N} c_n a_n^z$, where $\{a_1,\ldots,a_N\} = A$ and $c_n \in \{-1,0,1\}$. Choose the c_n not all 0 so that

$$|f(z)| \leq c_1 Nm(N)^{N^\alpha} \exp(-c_2 N/N^\alpha)$$

for some $c_1, c_2 > 0$ and all $z \in \underset{\sim}{Z}$ with $1 \leq z \leq N^{\alpha}$. Then assume that the result is false.)

*<u>8.2.4</u>. To what extent can the conclusion of Exercise 8.2.3 be improved?

In connection with the construction of auxiliary functions $f(z)$ it is of great interest to find entire functions of exponential type for which $f(z) \in \underset{\sim}{Z}$ when $z \in \underset{\sim}{Z}$ and $z \geq 0$. Exercises 8.2.5 and 8.2.6 (essentially the work of G. Pólya) help to classify such functions.

<u>8.2.5</u>. Let $f(z), g(z)$ be entire functions such that $|f(z)|, |g(z)| < a^{|z|}$ for $|z|$ sufficiently large, where $1 < a < e$. If $f(n) = g(n)$ for $n \in \underset{\sim}{Z}^+$, then $f(z) = g(z)$ for all $z \in \underset{\sim}{C}$. (Hint: Say $f(\alpha) - g(\alpha) \neq 0$ for some $\alpha \in (0,1)$ and $|z|$ is large. Then $h(z) = f(z + \alpha) - g(z + \alpha)$ satisfies $|h(z)| < b^{|z|}$ for some b such that $a < b < e$, while $h(0) \neq 0$. Hence by Lemma 4.2.1 (Jensen's theorem)

$$\log \frac{n^n}{(1 - \alpha)\cdots(n - \alpha)} \leq \frac{1}{2\pi}\int_0^{2\pi} \log\left|\frac{h(ne^{i\theta})}{h(0)}\right| d\theta$$

$$\leq \log \frac{b^n}{|h(0)|} .$$

By letting $n \to \infty$ it follows that $e \leq b$, which is a contradiction.)

<u>8.2.6</u>. If $f(x)$ is entire, $f(z) \in \underset{\sim}{Z}$ when z is a nonnegative integer, and

$$(*) \qquad 2^{-r} \max_{|z|=r} |f(z)| \to 0 \quad \text{as} \quad r \to \infty,$$

then $f(z)$ is a polynomial. (Hint: Let $I_n = \sum_{v=0}^{n}(-1)^v\binom{n}{v}f(n - v) \in \underset{\sim}{Z}$.

Then a partial fraction decomposition yields the integral representation

$$I_n = \frac{1}{2\pi i} \int_{|z|=2n} f(z)n![z(z - 1)\cdots(z - n)]^{-1}\, dz.$$

Set $z = 2ne^{i\theta}$. Then $|z - s| \geq 2n - s \cos\theta$ for s real, so for any $\epsilon > 0$ and n sufficiently large, (*) yields

$$|I_n| \leq \frac{1}{2\pi} \epsilon n! 2^{2n} \prod_{s=1}^{n} (2n - s)^{-1} \int_{-\pi}^{\pi} \prod_{s=1}^{n} [(2n - s)(2n - s\cos\theta)^{-1}]\, d\theta.$$

Now there are positive absolute constants c_1, c_2, c_3 such that

$$(1 - t)(1 - t\cos\theta) \leq e^{-c_1 t\theta^2} \qquad (0 \leq t \leq \tfrac{1}{2}),$$

$$\int_{-\pi}^{\pi} e^{-n\theta^2}\, d\theta \leq c_2 n^{-1/2},$$

$$n! 2^{2n} \prod_{s=1}^{n} (2n - s)^{-1} \leq c_3 n^{1/2}.$$

Thus $I_n = 0$ for n sufficiently large. Now apply Exercise 8.2.5 with a = 2 to show that f(z) is the polynomial

$$g(z) = \sum_{n=0}^{\infty} \frac{z(z - 1)\cdots(z - n + 1)I_n}{n!}.$$

For more details see Landau [51].)

It is harder to show that if (*) is replaced by

$$e^{-Ar} \max_{|z|=r} |f(z)| \to 0 \quad \text{as} \quad r \to \infty,$$

then $f(z) = P_0(z) + P_1(z)2^z$ for $A < 0.7588\ldots =$

$|\ell n(3/2 + 3^{1/2}i/2)|$, where $P_0(z), P_1(z) \in \underset{\sim}{Q}[z]$. Also, if (*) is replaced by

$$|f(z)| \leq c_1 e^{c_2|z|}$$

$$|f(x)| \leq \begin{cases} c_3 e^{ax} & (x \geq 0), \\ c_4 e^{-bx} & (x < 0), \end{cases}$$

$$|f(ix)| \leq c_5 \qquad (x \text{ real}),$$

where $c_1, \ldots, c_5 > 0$ are absolute constants and $e^a - e^{-b} < \sqrt{5}$, then $f(z) = P_0(z) + P_1(z)2^z + \cdots + P_m(z)m^z$, where $P_0(z), \ldots, P_m(z) \in \underset{\sim}{Q}[z]$. See Buck [16] and R. M. Robinson [92].

8.3. Hermite Interpolation

In the proof of Proposition 8.2.1 we were able to determine the behavior of f(z) when |z| was not too large from the values of $f^{(i)}(0)$ for $0 \leq i \leq M - 1$. In general, let $\varphi(z)$ be analytic. Given $\varphi^{(i)}(j)$ for $i = 0,1,\ldots,I - 1$ and $j = 0,1,\ldots,J - 1$, we wish to estimate $\varphi(z)$. If $J = 1$, the Taylor-series expansion of $\varphi(z)$ about $z = 0$ suggests writing

$$\varphi(z) = \varphi(0) + z\varphi'(0) + \frac{z^2\varphi''(0)}{2!} + \cdots + \frac{z^{m-1}\varphi^{(m-1)}(0)}{(m - 1)!} + E_m,$$

where $m \in \underset{\sim}{Z}^+$. Since $\varphi^{(i)}(0)/i! = (2\pi i)^{-1} \int_{\mathcal{C}} \varphi(\zeta)\zeta^{-i-1}\, d\zeta$, where $\mathcal{C}$ is a contour containing 0, we can get an integral representation of E_m by multiplying both sides of the identity

$$(8.3.1)\quad (\zeta - z)^{-1} = (\zeta^{-1} + z\zeta^{-2} + \cdots + z^{m-1}\zeta^{-m}) + (z/\zeta)^m(\zeta - z)^{-1}$$

by $\varphi(\zeta)$ and integrating them around a contour $\mathcal{C}$ containing both 0 and z. Thus

$$E_m = (2\pi i)^{-1} \int_{\mathcal{C}} \varphi(\zeta)\, \left(\frac{z}{\zeta}\right)^m (\zeta - z)^{-1}\, d\zeta.$$

In exactly the same way we see that if z is <u>outside</u> a disk of radius R about 0, then

$$z^i = (2\pi i)^{-1} \int_{|\zeta|=R} \zeta^i\, \left(\frac{z}{\zeta}\right)^m (z - \zeta)^{-1}\, d\zeta.$$

Hence

$$(8.3.2)\quad \varphi(z) = \sum_{i=0}^{m-1} \frac{\varphi^{(i)}(0)}{i!} \frac{1}{2\pi i} \int_{|\zeta|=R} \zeta^i\, \left(\frac{z}{\zeta}\right)^m (z - \zeta)^{-1}\, d\zeta + \frac{1}{2\pi i} \int_{\mathcal{C}} \varphi(\zeta)\, \left(\frac{z}{\zeta}\right)^m (\zeta - z)^{-1}\, d\zeta.$$

To get a similar result for the Taylor expansion of $\varphi(z)$ about $z = 1$ we would replace z and ζ by $z - 1$ and $\zeta - 1$ in (8.3.1), and the factor $[(z - 1)/(\zeta - 1)]^m$ would then occur on the right-hand side of (8.3.2). More generally, we have what Gelfond ([37], p. 35; see Exercise 8.3.8) calls the Hermite interpolation formula:

<u>Theorem 8.3.1</u>. Let the simple closed contour $\mathcal{C}$ contain z and $0,1,\ldots,J - 1$. If $\varphi(z)$ is analytic on and inside $\mathcal{C}$, and z lies outside closed disks of radius R about $0,1,\ldots,J - 1$, then

$$(8.3.3)\quad \varphi(z) = \sum_{j=0}^{J-1}\sum_{i=0}^{I-1} \frac{\varphi^{(i)}(j)}{i!}\,\frac{1}{2\pi i}\int_{|\zeta-j|=R} (\zeta - j)^{i}(z - \zeta)^{-1}$$

$$\cdot \prod_{v=0}^{J-1} \left(\frac{z - v}{\zeta - v}\right)^{I} d\zeta$$

$$+ \frac{1}{2\pi i}\int_{\mathcal{C}} \varphi(\zeta)(\zeta - z)^{-1} \prod_{v=0}^{J-1} \left(\frac{z - v}{\zeta - v}\right)^{I} d\zeta.$$

Proof. Apply the residue theorem to $(2\pi i)^{-1}\int_{\mathcal{C}} f(\zeta)\, d\zeta$, the final term of (8.3.3). At $\zeta = z$, the residue is $\varphi(z)$. To find the residue at $\zeta = j$, $0 \le j \le J - 1$, write

$$f(\zeta) = F(\zeta) + \sum_{i=1}^{I} a_i(\zeta - j)^{-i},$$

where $F(\zeta)$ is analytic at $\zeta = j$, and let

$$P_j(\zeta) = P(\zeta) = \prod_{v\neq j} \left(\frac{z - v}{\zeta - v}\right)^{I},$$

where $0 \le v \le J - 1$. Then

$$P(\zeta)\varphi(\zeta)(\zeta - z)^{-1} = \left(\frac{\zeta - j}{z - j}\right)^{I} f(\zeta),$$

so

$$\left[\frac{(z - j)^{I}}{(I - 1)!}\right] \left(\frac{d}{d\zeta}\right)^{I-1}[P(\zeta)(\zeta - z)^{-1}\varphi(\zeta)]\Big|_{\zeta=j} = a_1.$$

Hence by Leibniz's rule,

$$(8.3.4)\quad (2\pi i)^{-1}\int_{\mathcal{C}} f(\zeta)\,d\zeta = \varphi(z) + \sum_{j=0}^{J-1}\left[\frac{(z-j)^I}{(I-1)!}\right]$$

$$\cdot \sum_{i=0}^{I-1}\binom{I-1}{i}\left[P(\zeta)(\zeta - z)^{-1}\right]^{(I-1-i)}\Big|_{\zeta=j}\,\varphi^{(i)}(j).$$

Now

$$(8.3.5)\quad \left[P(\zeta)(\zeta - z)^{-1}\right]^{(I-1-i)}\Big|_{\zeta=j}$$

$$= (I - 1 - i)!(2\pi i)^{-1}\int_{|\zeta-j|=R}$$

$$\cdot P_j(\zeta)(\zeta - z)^{-1}(\zeta - j)^{-I+i}\,d\zeta,$$

so (8.3.3) follows from (8.3.5).

□

In this volume (8.3.3) will be used mostly to discuss $\varphi(z)$ for $|z| \geq J$ when something is known about the $\varphi^{(i)}(j)$ for $0 \leq i \leq I - 1$ and $0 \leq j \leq J - 1$; hence it could more aptly be called the Hermite extrapolation formula.

Exercises

In the following exercises we develop the rudiments of interpolation theory and obtain (Exercise 8.3.6) a Hermite formula slightly more general than that of Theorem 8.3.1. The form given by Gelfond in his treatise ([37], pp. 28-35) is that of Exercise 8.3.8.

8.3.1. Given a function $\varphi(z)$ and distinct complex numbers $z, z_0, z_1, \ldots,$ define $[z] = \varphi(z)$, $[z,z_0] = ([z] - [z_0])/(z - z_0)$,

$[z,z_0,z_1] = ([z,z_0] - [z_0,z_1])/(z - z_1),\ldots,$ and generally

$$(*) \qquad [z,z_0,\ldots,z_n] = \frac{[z,z_0,\ldots,z_{n-1}]-[z_0,z_1,\ldots,z_n]}{z - z_n} .$$

Show that

$$\varphi(z) = [z_0] + [z_0,z_1](z - z_0) + [z_0,z_1,z_2](z - z_0)(z - z_1)$$

$$+ \cdots + [z_0,\ldots,z_n](z - z_0)\cdots(z - z_{n-1})$$

$$+ [z,z_0,\ldots,z_n](z - z_0)\cdots(z - z_n)$$

by iterating (*) and use this to motivate the common approximation

$$\varphi(z) \sim \varphi(0) + \varphi'(0)z + \frac{\varphi''(0)z^2}{2!} + \cdots + \frac{\varphi^{(n)}(0)z^n}{n!} .$$

<u>8.3.2</u>. In the notation of Exercise 8.3.1 show that

$$[z_0,z_1,\ldots,z_n] = \sum_{j=0}^{n} \frac{\varphi(z_j)}{\prod_{i=0}^{n}{}' (z_j - z_i)}$$

when the z_i are distinct.

<u>8.3.3</u>. Let $z,z_1,\ldots,z_n$ be any complex numbers and H their convex hull. Say $\varphi(z)$ is holomorphic in an open set S containing H. Put $u_0(z) = \varphi(z)$ and

$$u_k(z) = \int_0^1 \int_0^{t_1} \cdots \int_0^{t_{k-1}} \varphi^{(k)}(\zeta_k)\, dt_k \cdots dt_1$$

for k = 1,2,...,n, where

$$\zeta_k = z_1 + (z_2 - z_1)t_1 + \cdots + (z - z_k)t_k.$$

Show that $\zeta_k \in H$ and

$$u_k(z) = \frac{u_{k-1}(z) - u_{k-1}(z_k)}{z - z_k}$$

for k = 1,2,...,n. (Hint: Integrate with respect to t_k.)

8.3.4. In Exercise 8.3.3 show that $u_k(z) = [z,z_1,\ldots,z_k]$ whenever the right side is defined.

In the following we redefine $[z,z_1,\ldots,z_k]$ as $u_k(z)$; by Exercise 8.3.4 this merely makes the symbol meaningful for all complex $z,z_1,\ldots,z_k$.

8.3.5. In Exercise 8.3.3 show that

$$[z_0,z_1,\ldots,z_n] = \frac{1}{2\pi i}\int_{\mathcal{C}} \frac{\varphi(\zeta)\, d\zeta}{(\zeta - z_0)\cdots(\zeta - z_n)} ,$$

where $\mathcal{C}$ is a contour in S containing $z_0,\ldots,z_n$ in its interior. (Hint: This follows from Exercise 8.3.2 and the Cauchy residue theorem if the z_i are distinct; in the general case use Exercise 8.3.3 and continuity.)

8.3.6. In Exercise 8.3.3 show that

$$\varphi(z) = \sum_{k=1}^{n} \sum_{\ell=0}^{p_k-1} \frac{\varphi^{(p_k-1-\ell)}(z_k)}{(p_k - 1 - \ell)!} \frac{1}{2\pi i}\int_{\mathcal{C}_k} \frac{1}{z - \zeta} (\zeta - z_k)^{p_k-\ell-1}$$

$$\cdot \prod_{k=1}^{n} \left(\frac{z - z_k}{\zeta - z_k}\right)^{p_k} d\zeta +$$

$$+ [z;z_1,\ldots,z_1;\ldots;z_n,\ldots,z_n] \prod_{k=1}^{n} (z - z_k)^{p_k},$$

where z_i is repeated p_i times in the last bracket. (Hint: Set $R = [z_0,\ldots,z_0;\ldots;z_n,\ldots,z_n]$, where z_i is repeated p_i times, $0 \leq i \leq n$. By Exercise 8.3.5,

$$R = \frac{1}{2\pi i} \int_{\mathcal{C}} \frac{\varphi(\zeta)\, d\zeta}{(\zeta - z_0)^{p_0} \cdots (\zeta - z_n)^{p_n}} = \sum_{k=0}^{n} I_k,$$

where

$$I_k = \frac{1}{2\pi i} \int_{\mathcal{C}_k} \frac{\varphi(\zeta)\, d\zeta}{(\zeta - z_0)^{p_0} \cdots (\zeta - z_n)^{p_n}}$$

for appropriate contours $\mathcal{C}_k$. Let $\Pi_0(\zeta) = \Pi_{i=0}^{n}(\zeta - z_i)^{p_i}$. Then by Cauchy's integral formula for the derivative of an analytic function,

$$I_k = (p_k - 1)!^{-1} D^{p_k - 1} \left[\frac{(\zeta - z_k)^{p_k} \varphi(\zeta)}{\Pi_0(\zeta)}\right]\Big|_{\zeta = z_k}$$

$$= \sum_{\ell=0}^{p_k - 1} \frac{\varphi^{(p_k - 1 - \ell)}(z_k)}{(p_k - 1 - \ell)!} \frac{1}{\ell!} D^{\ell}\left[\frac{(\zeta - z_k)^{p_k}}{\Pi_0(\zeta)}\right]\Big|_{\zeta = z_k}$$

$$= \sum_{\ell=0}^{p_k - 1} \frac{\varphi^{(p_k - 1 - \ell)}(z_k)}{(p_k - 1 - \ell)!} \frac{1}{2\pi i} \int_{\mathcal{C}_k} \frac{(\zeta - z_k)^{p_k}}{\Pi_0(\zeta)(\zeta - z_k)^{\ell+1}}\, d\zeta.$$

Set $p_0 = 1$ and $z_0 = z$. Then $\Pi_0(\zeta) = (\zeta - z)\Pi(\zeta)$, where

$$\Pi(\zeta) = \prod_{k=1}^{n} (\zeta - z_k)^{p_k},$$

and the result follows from

$$[z;z_1,\ldots,z_1;\ldots;z_n,\ldots,z_n] = \frac{\varphi(z)}{\Pi(z)}$$

$$+ \sum_{k=1}^{n} \sum_{\ell=0}^{p_k-1} \frac{\varphi^{(p_k-1-\ell)}(z_k)}{(p_k - 1 - \ell)!} \frac{1}{2\pi i} \int_{\mathcal{C}_k} \frac{(\zeta - z_k)^{p_k-\ell-1}}{\Pi_0(\zeta)} d\zeta.$$

8.3.7. Deduce Theorem 8.3.1 (and in fact a generalization of it) from Exercise 8.3.6.

8.3.8 (Gelfond's form). In Exercise 8.3.3 show that

$$\varphi(z) = \sum_{k=1}^{n} \sum_{\ell=0}^{p_k-1} \sum_{m=0}^{\ell} \frac{\varphi^{(p_k-1-\ell)}(z_k)}{(p_k - \ell - 1)!(\ell - m)!} D^{\ell-m}\left[\frac{(\zeta - z_k)^{p_k}}{\Pi(\zeta)}\right]\Big|_{\zeta=z_k}$$

$$\cdot \Pi(z)(z - z_k)^{-m-1} + \Pi(z)[z;z_1,\ldots,z_1;\ldots;z_n,\ldots,z_n].$$

(Hint: Instead of finding an integral representation for the $D^{\ell}[\;]$ of the hint of Exercise 8.3.6, expand it by Leibniz's rule into

$$\sum_{m=0}^{\ell} \binom{\ell}{m} D^{\ell-m}\left[\frac{(\zeta - z_k)^{p_k}}{\Pi(\zeta)}\right]\Big|_{\zeta=z_k} (-1)^m m!(z_k - z)^{-m-1}.)$$

8.3.9. Show that the first term in the formula of Exercise 8.3.8 for $\varphi(z)$ is a polynomial of degree at most $-1 + \sum_{i=1}^{n} p_i$ such that

$$\varphi^{(i)}(z_k) = P^{(i)}(z_k)$$

for $0 \le i \le p_k - 1$ and $1 \le k \le n$.

8.3.10. Show that if $\varphi(z)$ is holomorphic inside and on the simple closed contour $\mathcal{C}$, then

$$\varphi(z) = \sum_{k=0}^{n} [z_0,\ldots,z_k](z - z_0)\cdots(z - z_k) + R_n(z),$$

where

$$R_n(z) = \frac{1}{2\pi i}\int_{\mathcal{C}} \prod_{k=0}^{n} \frac{z - z_k}{\zeta - z_k}\, \frac{\varphi(\zeta)}{\zeta - z}\, d\zeta.$$

8.4. Application to Flat Entire Functions with Discrete Behavior at Integral Points

Propositions 8.4.1, 8.4.2, and 8.4.3 show how the behavior of $\varphi(z)$ in a fairly large real interval $[0,z]$ can be deduced in some cases from the behavior of $\varphi^{(i)}(j)$ for $i = 0,1,\ldots,I - 1$ and $j = 0,1,\ldots,J - 1$, where J is "relatively small" and I is "relatively large." This result is similar to Proposition 8.2.1, although both its hypotheses and conclusion are stronger. In each case hypothesis (iii) corresponds to the fact that an $f \in \mathfrak{F}$ was integral at the nonnegative integers, while something like (ii) was previously obtained (for $j = 0$ only) by the pigeonhole principle. The proof of Proposition 8.4.1 requires not only the more sophisticated extrapolation technique of Section 8.3 but also an inductive procedure that will be of great importance later. (Note: These propositions hint that data on $\varphi^{(i)}(j)$ as a function of i

are more valuable than data as a function of j. For another example, $\varphi(z)$ entire and $\varphi^{(i)}(0) = 0$ for all $i \in \mathbb{Z}$, $i \geq 0$, implies that $\varphi(z) \equiv 0$, while $\varphi(z) = \sin \pi z$ satisfies $\varphi^{(0)}(j) = 0$ for all $j \in \mathbb{Z}$. However, the symmetry is restored if $\varphi(z)$ is of exponential type and $\limsup t^{-1} \log|\varphi(\pm it)| < \pi$ as $t \to \infty$; in that case a theorem of F. Carlson's (see Hille [48], p. 64) tells us that $\varphi(z) = 0$ for $z \in \mathbb{Z}$ and $z \geq 0$ implies $\varphi(z) \equiv 0$.)

To estimate the products occurring in the Hermite formula, we need the following lemma:

<u>Lemma 8.4.1</u>. Say $|\zeta - j| = 1/2$, $0 \leq j \leq J - 1$, and $0 \leq z \leq S$, where j, J, S are integers and $\zeta \in \mathbb{C}$. Then $|\Pi| \leq 8 \cdot 2^{S+4J}$, where

$$\Pi = \prod_{v=0}^{J-1} \frac{z - v}{\zeta - v} .$$

<u>Proof</u>. Let N and D denote the numerator and denominator of Π, respectively. First say $0 \leq j - 1$ and $j + 1 \leq J - 1$. Then

$$|N| \leq S(S - 1)\cdots(S - (J - 1)) \text{ and } |D| \geq \frac{1}{2} \cdot \frac{1}{2} \cdot \frac{1}{2}(j - 1)! \cdot (J - j - 2)!,$$

so

$$|\Pi| \leq 8 \frac{S!}{(S - J)!J!} \frac{J!}{(J - 3)!} \frac{(J - 3)!}{(j - 1)!(J - j - 2)!}$$

$$\leq 8 \cdot 2^S J(J - 1)(J - 2)2^{J-3} \leq 8 \cdot 2^{S + 4J}$$

since $J + 1 \leq 2^J$. If $j = 0$, then $|D| \geq \frac{1}{2} \cdot \frac{1}{2}(J - 2)!$ and

$$|\Pi| \leq 4 \frac{S!}{(S-J)!J!} \frac{J!}{(J-2)!} \leq 4 \cdot 2^S J(J-1) \leq 8 \cdot 2^{S+4J};$$

the same estimate holds when $j = J - 1$. □

Corollary 8.4.1. If $0 \leq |z| \leq S$, then $|\Pi| \leq 8 \cdot 2^{2S+5J-1}$.

Proof. One modifies the previous estimate by a factor of

$$\frac{S(S+1)\cdots(S+J-1)}{S(S-1)\cdots(S-(J-1))} = \frac{(S+J-1)!}{(S-1)!J!} \frac{J!(S-J)!}{S!}$$

$$\leq 2^{S+J-1}.$$ □

The inductive procedure that we use to prove Proposition 8.4.1 is indicated schematically in Figure 6. At each stage we know that

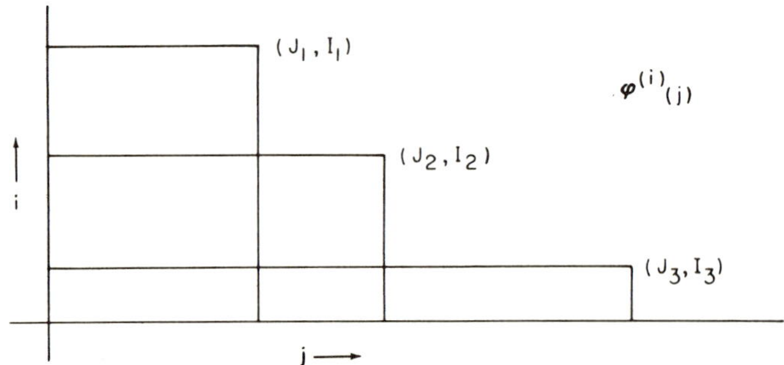

Figure 6

the $\varphi^{(i)}(j)$ are small in a certain $I_n \times J_n$ rectangle, and from this it is shown by extrapolation that they are also small in an $I_{n+1} \times J_{n+1}$ rectangle where $J_{n+1} > J_n$ but $I_{n+1} < I_n$.

Proposition 8.4.1. Let $A, I > 1$ and let i, j denote integers. Define $\mathfrak{F}$ to be the set of all entire functions $\varphi(z)$ such that, for

$0 \leq i \leq I$,

(i) $|\varphi^{(i)}(z)| \leq \exp\{A(|z| + 1)\}$.

(ii) $\varphi^{(i)}(0) = 0$.

(iii) If $|\varphi^{(i)}(j)| < \frac{1}{4}$, then $\varphi^{(i)}(j) = 0$.

If $\varphi \in \mathfrak{F}$, then

$$(*) \qquad \varphi(j) = 0 \text{ for } 0 \leq j \leq \frac{1}{20} \exp\left(\frac{1}{100} \ln^2 \frac{I}{200A}\right).$$

This is a special case of Proposition 8.4.2. In fact, the function on the right of (*) can be replaced by one that grows <u>exponentially</u> with I. This will be done in Proposition 8.4.3. However, it is best for the reader to learn the proof of the weaker proposition first. In fact, the techniques used in the proof of Proposition 8.4.2 are the same as those used to derive the results of A. Baker and N. I. Feldman, which appear later in this book in Chapters 9, 10 and 12. Proposition 8.4.3 may be omitted on a first reading.

<u>Definition 8.4.1</u>. Let $\psi = \{A,I,J,a,b,g,r\}$ be a set of positive real numbers with $A,I,J,a,b \geq 1$ and $1 + I^{-1}a \leq r$. Let i and j denote integers, and $\varphi(z)$ a function holomorphic for $|z| \leq r$.

<u>Proposition 8.4.2</u>. Let $\mathfrak{F} = \mathfrak{F}(\psi)$ be the set of all $\varphi(z)$ such that

(i) $|\varphi^{(i)}(z)| \leq \exp\{Ab(|z| + 1)\}$, $0 \leq i \leq I$, $|z| \leq r$.

(ii) $|\varphi^{(i)}(j)| \leq e^{-a}$, $0 \leq i \leq I$, $0 \leq j \leq J - 1$.

(iii) If $|\varphi^{(i)}(j)| \leq e^{-b}$, then $|\varphi^{(i)}(j)| \leq e^{-a}$,

provided $0 \leq i \leq I$ and $0 \leq j \leq g$.

If $\varphi \in \mathfrak{F}$, then $|\varphi(j)| \leq e^{-a}$ for

$$0 \leq j \leq J \min\{\frac{1}{20} \exp(\frac{1}{100} \ln^2 \frac{I}{200Ab}), \frac{1}{20}(\frac{ab}{10JI^2})^{10^{-4}}, \frac{g}{J}, \frac{r}{J}\}.$$

By setting $J = b = 1$ and $a = g = r = +\infty$, we obtain Proposition 8.4.1.

Proposition 8.4.2 is trivial unless

$$I > 2000Ab, \tag{8.4.1}$$

so we make this assumption. It is also trivial unless

$$a > 10\tau, \tag{8.4.2}$$

so we make this assumption as well. Let $[x]$ denote the greatest integer in x, $I_1 = [I]$, $J_1 = [J]$, and let i,j,s,t,u denote integers. The proof of Proposition 8.4.2 rests on the following lemma:

Lemma 8.4.2. Let (i), (8.4.1), and (8.4.2) hold, and let

$$|\varphi^{(i)}(j)| \leq e^{-a} \qquad (0 \leq i \leq I_u,\ 0 \leq j \leq J_u - 1), \tag{8.4.3}$$

where

(1) $400Ab \leq I_u \leq I_1$.

(2) $1 \leq J_1 \leq J_u \leq \frac{1}{10} \frac{ab}{I_1^2}$.

Then

$$|\varphi^{(t)}(j)| \leq e^{-2b} \quad \text{for} \quad 0 \leq t \leq [\tfrac{1}{2} I_u] \qquad (0 \leq j \leq S_u), \tag{8.4.4}$$

where $S_u = [I_u J_u/100Ab]$.

<u>Proof</u>. Set $I' = [\frac{1}{2} I_u]$, $J = J_u$, $S = S_u$, and $\rho = \rho_u = 2S_u$. Note that $S > 2J$. Also, (8.4.4) need be proved only for $J \leq j \leq S$. For $J \leq z \leq S$ and $0 \leq t \leq I'$, the Hermite formula yields

$$(8.4.5) \quad \varphi^{(t)}(z) = \sum_{j=0}^{J-1} \sum_{i=0}^{I'=1} \frac{\varphi^{(i+t)}(j)}{i!} \frac{1}{2\pi i} \int_{|\zeta - j| = \frac{1}{2}}$$

$$\cdot \frac{(\zeta - j)^i}{z - \zeta} \prod_{v=0}^{J-1} \left(\frac{z - v}{\zeta - v}\right)^{I'} d\zeta$$

$$+ \frac{1}{2\pi i} \int_{|\zeta|=\rho} \frac{\varphi^{(t)}(\zeta)}{\zeta - z} \prod_{v=0}^{J-1} \left(\frac{z - v}{\zeta - v}\right)^{I'} d\zeta = T_1 + T_2.$$

The Hermite formula (Theorem 8.3.1) and (i) are applicable because

$$\rho = 2S \leq \frac{1}{50} b^{-1} I_1 \cdot \frac{1}{10} \frac{ab}{I_1^2} = \frac{a}{500 I_1} < r$$

by Definition 8.4.1. Note that $(i + t) \leq I_u$ in (8.4.5). Hence by Lemma 8.4.1, (8.4.3) and $S > 2J$,

$$|T_1| \leq I'Je^{-a} \cdot \frac{1}{2} \cdot 2 \cdot 8^{I'} 2^{I'(S+4J)} \leq e^{-a} 2^{10I'J+I'S}$$

$$\leq e^{-a} 2^{6I'S} < e^{-\frac{1}{2} a}$$

since

$$6I'S \leq 6 \cdot \frac{1}{2} I_1 \cdot \frac{1}{100} b^{-1} I_1 \cdot \frac{1}{10} \frac{ab}{I_1^2} < \frac{1}{2} a.$$

Next,

$$|T_2| \le \frac{\rho}{\rho - S} e^{Ab(\rho+1)} \left(\frac{S}{\rho - J}\right)^{I'J} \le 2e^{Ab} e^{2AbS - (\ln \frac{3}{2}) I'J}.$$

Since $AbS \le I'J/25$ and $0.4 < \ln \frac{3}{2} < 0.41$, we have, by (1) and (2),

$$|T_2| \le \exp(1 + Ab - 8I'/25) \le \exp(2Ab - 32Ab)$$
$$= \exp(-30b).$$

Finally, by (8.4.2),

$$|\varphi^{(t)}(z)| \le |T_1| + |T_2| \le \exp(-5b) + \exp(-30b) \le \exp(-2b).$$

In particular this holds when $z = j \in \underset{\sim}{Z}$, and the lemma is proved. □

For the proof of Proposition 8.4.2, set $I_{u+1} = [\frac{1}{2} I_u]$ and $J_{u+1} = S_u$ for $u \ge 1$. Since $e^{-2b} < e^{-b}$, whenever the conclusion of the lemma is valid we have, by (iii), that

(8.4.6) $\quad |\varphi^{(i)}(j)| \le e^{-a} \quad$ for $\quad 0 \le i \le I_{u+1},\ 0 \le j \le J_{u+1} - 1.$

By (ii) we see that (8.4.6) is valid for $u = 0$. Thus (8.4.6) is valid for $0 \le u \le U$ by induction provided (1) and (2) hold for $u \le U$, and $J_{U+1} \le \max(g,r)$ so (i) and (iii) are applicable. We choose

$$U = \left[\min\left\{\frac{1}{10} \ln \frac{I}{200Ab}, \frac{1}{10^3} \frac{\ln(ab/10JI^2)}{\ln(I/100Ab)}\right\}\right]$$
$$= [\min\{U_1, U_2\}].$$

Since $U \le U_1$,

$$I_U \ge 4^{-U+1} I_1 \ge 2\cdot 4^{-U} I \ge 2I \exp\{-(\ell n\ 4)\cdot \frac{1}{10}\ \ell n \frac{I}{200Ab}\}$$

$$\ge 2I\cdot \frac{200Ab}{I}$$

and (1) follows. For (2), note that

$$\frac{4^{-u}I}{100Ab} \le \frac{1}{2}\ \frac{I_u}{100Ab} \le \frac{J_{u+1}}{J_u} \le \frac{I_u}{100Ab} \le \frac{2^{-u}I}{50Ab},$$

so

$$(8.4.7)\qquad 2^{-u^2}\left(\frac{I}{200Ab}\right)^u \le \frac{J_{u+1}}{J_1} \le 2^{-(u^2-u)/2}\left(\frac{I}{100Ab}\right)^u .$$

Since $U \le U_2$, we have

$$J_{U+1} \le J_1 \exp(U_2\,\ell n \frac{I}{100Ab}) = J_1 \exp(\frac{1}{10^3}\ \ell n \frac{ab}{10JI^2})$$

$$\le \frac{ab}{10I_1^2}.$$

The conclusion will now follow from the left-hand side of (8.4.7). If $U = [U_1]$, then $U \ge U_1 - 1$ and

$$J_{U+1} \ge \frac{1}{2}\ J \exp(\frac{1}{10}\ \ell n^2 \frac{I}{200Ab} - \ell n \frac{I}{200Ab} - \frac{\ell n\ 2}{100}\ \ell n^2 \frac{I}{200Ab})$$

$$\ge \frac{200Ab}{I}\cdot \frac{1}{2}\ J \exp(\frac{1}{100}\ \ell n^2 \frac{I}{200Ab})$$

$$\ge \frac{1}{20}\ J \exp(\frac{1}{100}\ \ell n^2 \frac{I}{200Ab}).$$

If $U \ne [U_1]$, then $U = [U_2]$ and $U_2 < U_1$. Also, $2\ \ell n(I/200Ab) > \ell n(I/100Ab)$. Thus

$$J_{U+1} \geq \frac{1}{2} J \exp\{\frac{1}{2} \cdot \frac{1}{10^3} \ln \frac{ab}{10JI^2} - \ln \frac{I}{200Ab} - \frac{\ln 2}{10^6} \frac{\ln^2(ab/10JI^2)}{\ln^2(I/100Ab)}\}$$

$$\geq \frac{100Ab}{I} \cdot J \exp\{\frac{1}{2} \cdot \frac{1}{10^3} \ln \frac{ab}{10JI^2} - \frac{\ln 2}{10^4} \ln \frac{ab}{10JI^2}\}$$

$$\geq \frac{1}{20} J(\frac{ab}{10JI^2})^{10^{-4}}.$$

This proves Proposition 8.4.2.

Essentially, we repeatedly ($u = 1,2,\dots$) extended our knowledge of $\varphi^{(i)}(j)$ as a function of j by a factor of $4^{-u}I/100Ab$ at the cost of halving the values of i for which we could estimate it. In other words, I_u decreased <u>exponentially</u> with u, the number of extrapolations. We now introduce a new procedure in which I_u decreases only <u>linearly</u> with u. This will give stronger results.

<u>Proposition 8.4.3</u>. The conclusion of Proposition 8.4.2 is valid for

$$0 \leq j \leq J \min\{\frac{1}{5} \exp(\frac{I}{2000Ab}), \frac{1}{5}(\frac{ab}{10JI^2})^{1/3}, \frac{g}{J}, \frac{r}{J}\}.$$

As before, we can assume that

$$I > 2000Ab \tag{8.4.8}$$

and

$$a > 10b. \tag{8.4.9}$$

Let $[x]$ denote the greatest integer in x, $I_1 = [I]$, $J_1 = [J]$, and let i,j,s,t,u denote integers. Proposition 8.4.3 rests on the following lemma:

Lemma 8.4.3. Let (i), (8.4.8), and (8.4.9) hold, and let

(8.4.10) $$|\varphi^{(i)}(j)| \le e^{-a} \qquad (0 \le i \le I_u,\ 0 \le j \le J_u - 1),$$

where

(1) $0 \le I_u \le I_1$.

(2) $1 \le J_1 \le J_u \le ab/10I_1^2$.

Then

(8.4.11) $$|\varphi^{(t)}(j)| \le e^{-2b} \quad \text{for} \quad 0 \le t \le [I_u - 650Ab]$$

and $0 \le j \le [2.5J_u] + 1$.

Proof. We first show (8.4.11) for

(8.4.12) $$0 \le t \le [\tfrac{1}{2}(I_u + s)] + 1, \qquad J_u \le j \le S_u = S_u(s)$$

where

(8.4.13) $$0 \le s \le I_u - 1250Ab \quad \text{and} \quad S_u = [\frac{\frac{1}{2}(I_u - s)J_u}{100Ab}].$$

Then (8.4.11) follows by letting $s = s_0 = [I_u - 1250Ab]$. Note that

(8.4.14) $$S_u(s) \ge S_u(s_0) \ge 2.5J_u > 2J_u.$$

For the rest of the proof set $I' = [\frac{1}{2}(I_u - s)]$, $J = J_u$, $S = S_u$, and $\rho = \rho_u = 2S_u$. For $J \le z \le S$ and $0 \le t \le I'$, the Hermite formula (Theorem 8.3.1) is applicable because

$$\rho = 2S \le \frac{I_1 J}{100Ab} \le \frac{I_1 ab}{100Ab}\ 10I_1^2 \le r$$

by Definition 8.4.1. Thus (8.4.5) holds, and clearly $i + t \le I_u$

in (8.4.5). Hence by Lemma 8.4.1, (8.4.10), and $S > 2J$, we have

$$|T_1| < \exp(-a/2)$$

as in Lemma 8.4.2 since

$$6I'S \le 6I_1 \frac{I_1}{200Ab} \frac{ab}{10I_1^2} < \frac{1}{2} a.$$

Again, as in Lemma 8.4.2,

$$|T_2| \le \exp(1 + Ab + 2AbS - 0.4I'J).$$

Since $I' \ge 600Ab > 1$, we have

$$2AbS \le 2Ab \frac{2I'J}{100Ab} = \frac{I'J}{25}.$$

Therefore $(J \ge 1)$

$$|T_2| \le \exp(2Ab - 0.36I') \le \exp(2Ab - 216Ab) < \exp(-200b).$$

Finally, by (8.4.9),

$$|\varphi^{(t)}(z)| \le |T_1| + |T_2| \le e^{-5b} + e^{-200b} \le e^{-2b}.$$

In particular this holds when $z = j \in \underset{\sim}{Z}$, and the lemma is proved. □

For the proof of Proposition 8.4.3, set $I_{u+1} = [I_u - 650Ab]$ and $J_{u+1} = [2.5J_u] + 1$ for $u \ge 1$. Since $e^{-2b} < e^{-b}$, whenever the conclusion of the lemma is valid, we expect by (iii) that

$$|\varphi^{(i)}(j)| \le e^{-a} \quad \text{for} \quad 0 \le i \le I_{u+1},\ 0 \le j \le J_{u+1} - 1. \tag{8.4.15}$$

By (ii), we see that (8.4.15) is valid for $u = 0$. Thus (8.4.15) is valid for $0 \le u \le U$ by induction provided (1) and (2) hold for $u \le U$, and $J_{U+1} \le \max(g,r)$ so (i) and (iii) are applicable. We choose $U = [\min(U_1, U_2)]$, where

$$U_1 = \frac{I}{1400Ab} \quad \text{and} \quad U_2 = \frac{1}{2} \ell n \frac{ab}{10JI^2}.$$

The inequality $I_{u+1} \ge I_u - 700Ab$ implies that $I_U \ge I_1 - \frac{1}{2} I \ge 0$, so (1) follows. For (2), note that

$$2.5 \le \frac{J_{u+1}}{J_u} \le 3.5,$$

so

$$(8.4.16) \qquad (2.5)^U \le \frac{J_{U+1}}{J_1} \le (3.5)^U \le \left(\frac{ab}{10JI^2}\right)^{\frac{1}{2} \ell n(3.5)}.$$

Thus

$$J_{U+1} \le J_1\left(\frac{ab}{10JI_1^2}\right),$$

and (2) follows. If $U = [U_1]$, then

$$J_{U+1} \ge \frac{1}{2} J \exp\{(\ell n\ 2.5)(\frac{I}{1400Ab} - 1)\}$$

$$\ge \frac{1}{5} J \exp(\frac{I}{2000Ab}).$$

If $U = [U_2]$, then

$$J_{U+1} \ge \frac{1}{2} J \exp\{(\ell n\ 2.5)(\frac{1}{2} \ell n \frac{ab}{10JI^2} - 1)\}$$

$$\ge \frac{1}{5} J\left(\frac{ab}{10JI^2}\right)^{1/3}.$$

This proves Proposition 8.4.3.

A peculiarity of Propositions 8.4.2 and 8.4.3 is that "beyond some point" knowing that I is extremely large does not strengthen the conclusion. In applications it might be advisable to replace a given I by a smaller one.

To see that the extrapolation procedure used in the proof of Proposition 8.4.3 is optimal in some sense, take $\lambda(x) = x$ in Exercise 8.4.3. See Ramachandra [86] for an application of a similar extrapolation procedure.

Exercises

8.4.1. Show that the conclusion (*) of Proposition 8.4.1 can be replaced by

$$\text{(i)} \quad \varphi(j) = 0 \quad \text{for} \quad 0 \le j \le \frac{1}{5} \exp(I/2000A)$$

or

$$\text{(ii)} \quad \varphi^{(t)}(j) = 0 \quad \text{for} \quad 0 \le j \le \frac{1}{5} \exp(I/4000A),$$

t being an integer with $0 \le t \le I/2$.

In the next two exercises let $\lambda(x)$ be an increasing positive function such that $\lambda'(1) > 0$ and $\lambda''(x) \le 0$. Let μ be the inverse function of λ; that is, $\mu[\lambda(x)] = \lambda[\mu(x)] = x$.

8.4.2. Let $x \ge A > 0$ and define

$$\theta(x,A) = \min \sum_{n=1}^{N-1} \mu\left(\frac{A_{n+1}}{A_n}\right),$$

where the minimum is over all integers $N \ge 2$ and all real N-tuples $(A_1,\dots,A_N)$ such that $A = A_1 \le A_2 \le \cdots \le A_{N-1} \le A_n = x$. Show that

$$c_1(\mu)\ \ell n\ \frac{x}{A} \le \theta(x,A) \le c_2(\mu)\ \ell n\ \frac{x}{A},$$

where c_1, c_2 depend only on μ. (Hint: For the upper bound let $A_n = x^{n/N}A^{(N-n)/N}$. For the lower bound use the convexity of μ and the inequality of the arithmetic-geometric means.)

8.4.3. Let S be the smallest subset of $[0,\infty] \times [0,\infty]$ such that

(i) $[0,J] \times [0,I] \subseteq S$ for some $I,J > 0$.

(ii) If $[0,A] \times [t,B] \subseteq S$, then

$$[0, A\lambda(B - t)] \times [t, \tfrac{1}{2}(B + t)] \subseteq S.$$

Show that

$$Je^{c_3(\lambda)I} \le \max\{x \mid (x,y) \in S\} \le Je^{c_4(\lambda)I},$$

where c_3, c_4 depend only on λ. (Hint: Use Exercise 8.4.2.)

If we only use condition (ii) of Exercise 8.4.3 in the special case where $t = 0$, we shall have the extrapolation procedure of Proposition 8.4.2 rather than of Proposition 8.4.3.

8.5. Some Nonvanishing Determinants

In Sections 8.2, 8.3, and 8.4 we gave methods for proving that an $f(z)$ of the form (8.1.2) has a large number of integral zeros. Now we shall examine some cases where (8.1.3) is satisfied. If $f_n(z) = z^n$, then (8.1.3) is clearly satisfied since the determinant is Vandermondian [alternatively $f(z)$ has at most N zeros since it is a polynomial of degree N]. Condition (8.1.3) is also valid for $f_n(z) = n^z$ since transposing a matrix does not affect its determinant. The results of this section are a sort of mixture of these two cases.

Lemma 8.5.1. For distinct nonzero complex numbers $\omega_1,\ldots,\omega_\ell$, let $p,q \in \underset{\sim}{Z}^+$ and $D = D_q(\omega_1,\ldots,\omega_p) = \det(z^k\omega_\ell^z)$, where $0 \le k \le q-1$, $1 \le \ell \le p$, $0 \le z \le pq-1$, and the elements of D are ordered as in Figure 7. Then $D \ne 0$. In fact,

$$(8.5.1) \qquad D = [\det(z^k)]^p \prod_{\ell=1}^{p} \omega_\ell^{\binom{q}{2}} \prod_{\ell=2}^{p} [\prod_{\lambda<\ell} (\omega_\ell - \omega_\lambda)^{q^2}],$$

where $0 \le z, k \le q-1$ in $\det(z^k)$.

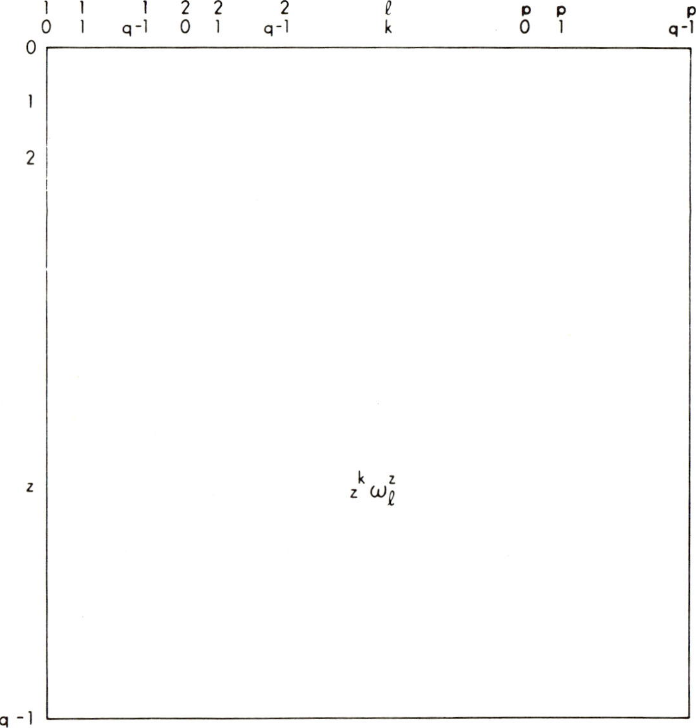

Figure 7

Note: If $q = 1$, this becomes the Vandermonde identity. For $p = q = 3$, we find that the determinant of Figure 8 equals $2(\omega_1\omega_2\omega_3)^3(\omega_2 - \omega_1)^9(\omega_3 - \omega_1)^9(\omega_3 - \omega_2)^9$.

$$
\left|\begin{array}{ccc:ccc:ccc}
1 & 0 & 0 & 1 & 0 & 0 & 1 & 0 & 0 \\
\omega_1 & \omega_1 & \omega_1 & \omega_2 & \omega_2 & \omega_2 & \omega_3 & \omega_3 & \omega_3 \\
\omega_1^2 & 2\omega_1^2 & 4\omega_1^2 & \omega_2^2 & 2\omega_2^2 & 4\omega_2^2 & \omega_3^2 & 2\omega_3^2 & 4\omega_3^2 \\
\hdashline
\omega_1^3 & 3\omega_1^3 & 9\omega_1^3 & & & & \omega_3^3 & 3\omega_3^3 & 9\omega_3^3 \\
\omega_1^4 & 4\omega_1^4 & 16\omega_1^4 & & & & \omega_3^4 & 4\omega_3^4 & 16\omega_3^4 \\
\omega_1^5 & 5\omega_1^5 & 25\omega_1^5 & & & & \omega_3^5 & 5\omega_3^5 & 25\omega_3^5 \\
\hdashline
 & & & & & & \omega_3^6 & 6\omega_3^6 & 36\omega_3^6 \\
 & & & & & & \omega_3^7 & 7\omega_3^7 & 49\omega_3^7 \\
 & & & & & & \omega_3^8 & 8\omega_3^8 & 64\omega_3^8
\end{array}\right|
$$

Figure 8

Proof. Write $D(\omega) = D_q(\omega_1,\ldots,\omega_{p-1},\omega)$. By considering the diagonal of the lower-right-hand $q \times q$ square of D (see Fig. 9), we find that

$$d_\omega[D(\omega)] = (pq - 1) + (pq - 2) + \cdots + (pq - q)$$

$$= pq^2 - \frac{1}{2}q(q + 1).$$

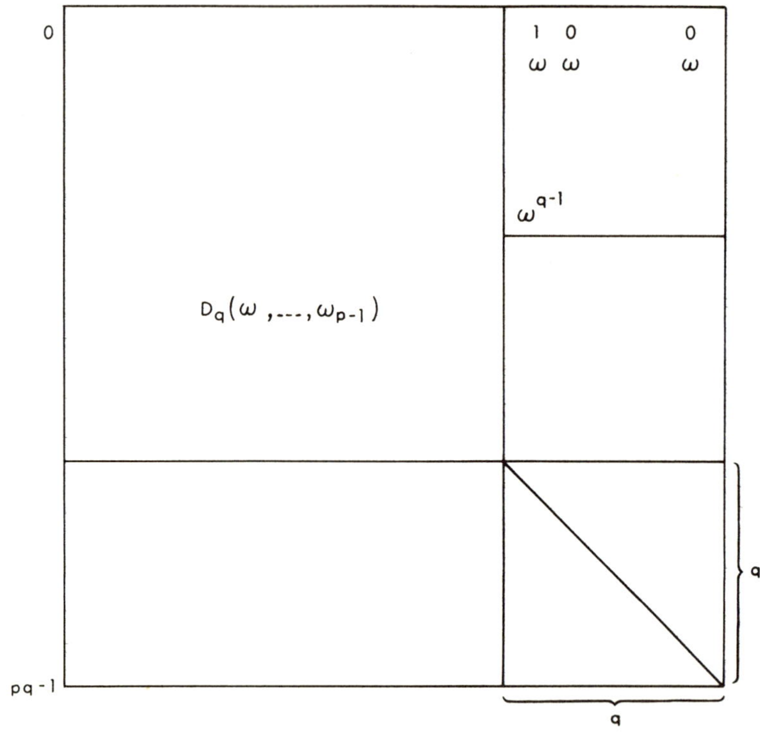

Figure 9

It is also clear from the nature of the last q columns of $D(\omega)$ that

$$\omega^{\binom{q}{2}} \mid D(\omega).$$

Thus, if we can show that $\Pi_{\ell=1}^{p-1}(\omega - \omega_\ell)^{q^2} \mid D(\omega)$, it will follow that

$$D(\omega) = \Delta_p \omega^{\binom{q}{2}} \prod_{\ell=1}^{p-1} (\omega - \omega_\ell)^{q^2}, \tag{8.5.2}$$

where Δ_p is independent of ω. This is equivalent to showing that

$$D^{(i)}(\omega)\big|_{\omega=\omega_\ell} = 0 \qquad (0 \le i \le q^2 - 1). \tag{8.5.3}$$

Now

$$D^i(\omega) = \sum_{i_1+\cdots+i_q = 1} D[i_1,\ldots,i_q](\omega),$$

where $D[i_1,\ldots,i_q](\omega)$ denotes $D(\omega)$ with the $(pq - q + j)^{th}$ column $(1 \le j \le q)$ differentiated i_j times. The zth row of this determinant has the form

$$\begin{aligned} &\ldots,\omega_\ell^z, z\omega_\ell^z,\ldots,z^{q-1}\omega_\ell^z,\ldots\ \ldots, \\ &\quad z(z-1)\cdots(z-i_1+1)\omega^{z-i_1},\ldots, \\ &\qquad z^{q-1}\cdot z(z-1)\cdots(z-i_q+1)\cdot\omega^{z-i_q}. \end{aligned} \tag{8.5.4}$$

On substituting ω_ℓ for ω and multiplying the $(pq - q + j)^{th}$ column by $\omega_\ell^{i_j}$, we find that the $2q$ elements in the zth row of

$$\omega_\ell^{i_1+\cdots+i_q} D^{(i)}(\omega_\ell)$$

that involve ω_ℓ have the forms $F_m(z)\omega_\ell^z$ respectively, $m = 1,\ldots,2q$, where $F_m(x) \in \underset{\sim}{Z}[x]$. If the $F_m(x)$ were linearly independent, then

$$\sum_{m=1}^{2q} d[F_m(x)] \ge \sum_{m=1}^{2q} (m - 1) = q(2q - 1).$$

However, from (8.5.4) we have

$$\sum_{m=1}^{2q} d[F_m(x)] = \frac{1}{2} q(q - 1) + \frac{1}{2} q(q - 1) + i_1 + \cdots + i_q$$

$$< q(q - 1) + q^2 = q(2q - 1)$$

when $0 \leq i \leq q^2 - 1$, so (8.5.3) follows. From (8.5.1) we see that Δ_p is the lead coefficient of $D(\omega)$ considered as a polynomial in ω. Hence (see Fig. 9)

$$\Delta_p = \det(z^k) D_q(\omega_1, \ldots, \omega_{p-1}), \tag{8.5.5}$$

where $0 \leq k \leq q - 1$ and $pq - q \leq z \leq pq - 1$. Now by the formula for a Vandermondian determinant (or by direct expansion) $\det[(z + c)^k] = \det(z^k)$ for any constant c. Hence we can assume that $0 \leq z \leq q - 1$ in (8.5.5). Thus by (8.5.5) and (8.5.2),

$$D_q(\omega_1, \ldots, \omega_p) = \det(z^k) \omega_p^{\binom{q}{2}} \prod_{\ell=1}^{p-1} (\omega_p - \omega_\ell)^{q^2} \cdot D_q(\omega_1, \ldots, \omega_{p-1}), \tag{8.5.6}$$

and (8.5.1) follows from a p-fold iteration of (8.5.6).

□

<u>Lemma 8.5.2</u>. Let $k, \ell, z; p, q; \omega_1, \ldots, \omega_p$ have the same ranges as in Lemma 8.5.1. Let $P_0(z), \ldots, P_{q-1}(z) \in \underset{\sim}{C}[z]$ be linearly independent over $\underset{\sim}{C}$ and let $d[P_i(z)] \leq q - 1$. Then for $t \in \underset{\sim}{C}$ and $t \neq 0$,

$$D = \det(P_k(tz)\omega_\ell^z) \neq 0.$$

Proof. Clearly the $\{P_k(tz)\}$ are linearly independent if and only if the $\{P_k(z)\}$ are linearly independent. By a p-fold application of the elementary column operations that transform the $1 \times q$ matrix

$$(P_0(tz),\ldots,P_{q-1}(tz))$$

into

$$(1,\ldots,z^{q-1})$$

we can transform D into $\det(z^k\omega_\ell^z)$, and this does not vanish by Lemma 8.5.1.

□

The results of this section can be found in Feldman [33], [34]; see also Lang [54], pp. 65-68.

Exercises

8.5.1. If $\cos\phi_1,\ldots,\cos\phi_m$ are distinct, show that $(0 \le i,j \le m)$ we have $\det(\cos i\phi_j) \neq 0$. (Hint: First show that $\cos i\phi = 2^{i-1}\cos^i\phi + \Sigma_{k=0}^{i-1} a_k \cos^k\phi$. This determinant occurs in Littlewood [61], p. 219.)

8.5.2. Show that 8.5.1 does not hold if $\cos i\phi_j$ is replaced by $\sin i\phi_j$. (Hint: Find a numerical counterexample for $m = 2$.)

8.5.3. Let $\omega_1,\ldots,\omega_m$ be distinct complex numbers, let $n_1,\ldots,n_m$ be integers, and let $N = \Sigma_{i=1}^m (n_i + 1)$. Let δ_{ij} be the Kronecker delta (1 if $i = j$ and 0 otherwise). Say there are polynomials $P_{ij}(x)$, where $1 \le i,j \le m$, such that

(i) $d[P_{ij}(x)] = n_i + \delta_{ij}$.

(ii) $R_i(x) \equiv \Sigma_{j=1}^{m} P_{ij}(x)e^{\omega_j x} = \Sigma_{k=N}^{\infty} c_{ik}x^k.$

Show that $D(x) \equiv \det P_{ij}(x) \neq 0$ if $x \neq 0$. (Hint: Clearly $d[D(x)] = N$. On the other hand, if we solve for $e^{\omega_1 x}D(x)$ by Cramer's rule, we find that it has a zero of order N at 0.)

It can be shown that polynomials $P_{ij}(x)$ satisfying the conditions of Exercise 8.5.3 do exist and that they are uniquely determined up to constant factors. The nonvanishing of determinants like D(x) can be used (see Siegel [110], p. 19) to prove the famous Lindemann-Weierstrass theorem, that if $\alpha_1,\ldots,\alpha_m$ are algebraic numbers that are linearly independent over $\mathbb{Q}$, then $e^{\alpha_1},\ldots,e^{\alpha_n}$ are <u>algebraically</u> independent over $\mathbb{Q}$. Mahler [65] further refined this method to prove the following quantitative version of this theorem: Say $K = \mathbb{Q}(\alpha_1,\ldots,\alpha_m)$, $n = d(K/\mathbb{Q})$, and $P = P(x_1,\ldots,x_m) \in \mathbb{Z}[x_1,\ldots,x_m]$. Then

$$|P(e^{\alpha_1},\ldots,e^{\alpha_m})| \geq H^{-cn_1\cdots n_m},$$

where $H = H(P)$, $n_i = d_{x_i}(P)$, and c is a positive constant depending only on n and m. The nonvanishing of determinants plays a central role in attempts to prove theorems of this sort for functions other than exp(x). Siegel [110] proved such theorems for certain "E-functions"; his methods were later extended by Shidlovskii [106].

8.6. Some Auxiliary Functions

A large number of classical transcendence theorems are special cases of a theorem of A. Baker's, which asserts that if $\log \alpha_1, \ldots, \log \alpha_m$ are linearly independent over $\underset{\sim}{Q}$, then $1, \log \alpha_1, \ldots, \log \alpha_m$ are linearly independent over $\underset{\sim}{A}$. In fact, by arguments following the pattern of Section 8.1, Baker (and later N. I. Feldman) gave explicit positive lower bounds for

$$|\beta_0 + \beta_1 \log \alpha_1 + \cdots + \beta_m \log \alpha_m| \qquad (\beta_i \in A)$$

in terms of α_i and β_i. The remainder of this volume will study these results and their implications.

To motivate the auxiliary functions used by Baker and Feldman, let us assume for simplicity that $\alpha_1, \ldots, \alpha_m$ are multiplicatively independent algebraic integers of degrees $\Delta_1, \ldots, \Delta_m$, respectively. (This is almost, but not quite, the same as having $\log \alpha_1, \ldots, \log \alpha_m$ linearly independent over $\underset{\sim}{Q}$; the fact that log z is multivalued will require careful consideration throughout.) Let

$$(8.6.1) \qquad g(z) = \sum_{e_m=0}^{E-1} \cdots \sum_{e_1=0}^{E-1} (\alpha_1^{e_1} \cdots \alpha_m^{e_m})^z C_{e_1 \cdots e_m}$$

for $z \in \underset{\sim}{C}$, where the $C_{e_1 \cdots e_m}$ are integers not all zero to be determined later. We can write

$$g(z) = \sum_{n=0}^{E-1} c_n f_n(z)$$

by ordering the terms lexicographically with $(e_1, \ldots, e_m)$. Now $D = \det[f_n(z)] \neq 0$, where $0 \leq n, z \leq E^m - 1$ $(z \in \underset{\sim}{Z})$, since D is

Vandermondian and the α_i are multiplicatively independent. Hence $g(z)$ can have at most $E^m - 1$ consecutive integral zeros beginning with 0. By the "theorem on the primitive element" (see van der Waerden [123], pp. 126 and 127) we can write

$$\underset{\sim}{K} = \underset{\sim}{Q}(\alpha_1,\ldots,\alpha_m) = \underset{\sim}{Q}(\theta),$$

where $d(\underset{\sim}{K}) = d(\theta)$. Put $n = d(\theta)$ and let σ_i $(1 \le i \le n)$ be the isomorphism of $\underset{\sim}{C}$ fixing $\underset{\sim}{Q}$ that takes θ into its ith conjugate θ_i; that is, $\sigma_i(\theta) = \theta_i$, $1 \le i \le n$. For $z \in \underset{\sim}{Z}$, $z \ge 0$, write $g_i(z) = \sigma_i[g(z)]$ and define

$$G(z) = (a_1^{\Delta_1}\cdots a_m^{\Delta_m})^z \prod_{i=1}^{n} g_i(z) , \tag{8.6.2}$$

where a_i is the lead coefficient of α_i. Clearly $G(z)$ is an integer, and since it vanishes only when $g(z) = 0$, it too can have at most $E^m - 1$ consecutive integral zeros beginning with 0. However, if certain false assumptions are made concerning the α_i, the methods of Sections 8.1, 8.2, and 8.3 enable one to prove the existence of <u>more</u> than $E^m - 1$ consecutive zeros for $G(z)$.

More generally one can let

$$g(z) = \sum_{k=0}^{q-1} \sum_{e_m=0}^{E-1} \cdots \sum_{e_1=0}^{E-1} P_k(tz)(\alpha_1^{e_1}\cdots\alpha_m^{e_m})^z C_{ke_1\cdots e_m}$$

with $P_k(tz)$ as in Lemma 8.5.2; by the conclusion of that lemma $g(z)$ can have at most $qE^m - 1$ consecutive integral zeros beginning with 0. In order for the corresponding $G(z)$ to be an integer, we require $P_k(tz) \in \underset{\sim}{Z}[z]$ for $0 \leq k \leq q - 1$. A possible choice is $P_k(tz) = P_k(z) = z^k$. However, to apply the methods of Sections 8.1, 8.2, and 8.3, we would like our auxiliary function together with its first s derivatives to be as small as possible. For this purpose Feldman chose

$$P_k(z) = (z + k + 1)(z + k + 2)\cdots(z + k + q - 1).$$

The idea is that we can later divide $g(z), g^{(1)}(z), \ldots, g^{(s)}(z)$ by $\gamma(q,s,z)$, the greatest common divisor of all integers in

$$\{(z + k + 1)^{\epsilon_1}(z + k + 2)^{\epsilon_2}\cdots(z + k + q - 1)^{\epsilon_{q-1}} \mid \epsilon_i = 0 \text{ or } 1,$$

$$\Sigma\, \epsilon_i \geq q - 1 - s, \text{ and } 0 \leq k \leq q - 1\}.$$

This leaves our auxiliary function with integral coefficients, but makes it quite a bit smaller (see Section 8.7). Of course we have to check that the determinant $D \neq 0$ here. It suffices by Lemma 8.5.2 to show that the $P_k(z)$ with $0 \leq k \leq q - 1$ are linearly independent. This follows from the following lemma:

Lemma 8.6.1. Let $P(x) = a_n x^n + \cdots + a_0$, where $n \geq 1$ and $a_n \neq 0$. Then, for any $h \neq 0$, the polynomials $P(x), P(x + h), \ldots, P(x + nh)$ are linearly independent.

Proof. This is clear if $n = 1$, so assume that it is true for $n - 1$. Let $Q(x) = P(x + h) - P(x)$. Then, for $0 \leq m \leq n - 1$, the polynomial $Q(x + m\alpha)$ is of degree exactly $n - 1$ with lead coefficient $nha_n \neq 0$. Hence these n polynomials are linearly independent by the induction hypothesis, and since $P(x)$ is of degree n, the result quickly follows. □

8.7. Distribution of Primes

Let $\pi(x)$ denote the number of primes p such that $p \le x$. The famous prime-number theorem asserts that

$$\lim_{x \to \infty} \pi(x) \ln x/x = 1,$$

but we only need the older (and much easier to prove!) result of Chebychev's that there exist effectively computable positive constants c_1, c_2 such that

$$c_1 x/\ln x < \pi(x) < c_2 x/\ln x. \tag{8.7.1}$$

For a proof of (8.7.1) see, for example, LeVeque [58], pp. 105-107. From (8.7.1) we deduce

$$\sum_{p \le n} \ln p \le (\ln n) \sum_{p \le n} 1 < c_2 n = O(n). \tag{8.7.2}$$

It also follows that

$$\sum_{p \le n} p^{-1} \ln p = \ln n + O(1). \tag{8.7.3}$$

To see this, let $[x]$ denote the greatest integer in x. Then $p^e \| n$, where

$$e = \sum_{i=1}^{\infty} [\frac{n}{p^i}].$$

Hence

$$\ln n! = \sum_{p\le n} \left(\sum_{i=1}^{\infty} \left[\frac{n}{p^i}\right]\right) \ln p$$

$$= \sum_{p\le n} \left[\frac{n}{p}\right] \ln p + O\left[n \sum_{p=2}^{\infty} \frac{\ln p}{p(p-1)}\right]$$

$$= n \sum_{p\le n} p^{-1} \ln p + O\left(\sum_{p\le n} \ln p\right) + O(n),$$

where we have used $x - 1 \le [x] \le x$. Now (8.7.3) follows from (8.7.2) and the simple estimate $\ln n! = n \ln n + O(n)$. Note that all these statements involving O's can be replaced by statements involving effectively computable positive constants.

The following lemmas from Feldman [34] will be useful later:

<u>Lemma 8.7.1</u>. Let $x,y,a \in \underset{\sim}{Z}^+$ and $y = 2^i$ for some nonnegative integer i. Then

$$(8.7.4) \qquad (x + y)(x + 2y)\cdots(x + ay)2^a/a! \in \underset{\sim}{Z}^+.$$

<u>Proof</u>. Let p be a prime, so $p^e \| a!$, where

$$e = \left[\frac{a}{p}\right] + \left[\frac{a}{p^2}\right] + \cdots \le \frac{a}{p-1} .$$

We must show that p^e divides the numerator N of the left-hand side of (8.7.4). If p = 2, this is clear. For $p > 2$, we must show that $f \ge 3$, where $p^f \| N$. For every $r \in \underset{\sim}{Z}^+$, the numerator N contains the (possibly empty) product

$$\prod_{i=0}^{I} \prod_{j=0}^{J} \{x + (ip^r + j)y\},$$

where $I = [a/p^r] - 1$ and $J = p^r - 1$. For i fixed and $0 \leq j \leq J$, the $x + (ip^r + j)y$ are a set of p^r integers such that no two are congruent modulo p^r. Hence one of them is congruent to 0 modulo p^r. It follows that $[a/p^r]$ factors in the numerator are divisible by p^r, so

$$f \geq \sum_{r=1}^{\infty} [\frac{a}{p^r}] = e.$$

□

Lemma 8.7.2. Let a,b,s be nonnegative integers with $a \geq 2(s + 1)$. Set $g_\pi(x) = \Pi_{i=1}^{a}(x + i)$ and

$$G = G(a,b,x) = \{g_\pi(x + j) | 0 \leq j \leq b\},$$

so G is a set of products having exactly a factors. Let $G_s = G(a,b,x;s)$ be the set of all products that can be obtained from elements of G by deleting at most s factors. Then if x is a nonnegative integer,

$$(8.7.5) \quad \gamma = \gamma(a,b,x;s) \geq a^a \exp(-a\{\ln(s + 1) + c \frac{\ln(x + a + b)}{\ln(a/(s + 1))}\}),$$

where γ is the greatest common divisor of G_s and c is an effectively computable positive constant.

Proof. Let p denote a prime. Then, for all c with $1 \leq c \leq a + b$, we have $p^e \| (x + c)$, where $e \leq \ln(x + a + b)/\ln p$. On the other hand, $p^f | g_\pi(x + j)$ for $0 \leq j \leq b$, where $f = [a/p]$. Hence

$$\ln \gamma \geq \sum_p = \sum_p ([\frac{a}{p}] - s \frac{\ln(x + a + b)}{\ln p}) \ln p,$$

where the sum is extended over any set of primes whatsoever. In particular,

$$\ln \gamma \geq \sum_{p \leq t}, \qquad \text{where } t = \frac{a}{s+1},$$

so

$$\begin{aligned}\ln \gamma &\geq a \sum_{p \leq t} p^{-1} \ln p - \sum_{p \leq t} \ln p - s\pi(t)\ \ln(x + a + b) \\ &\geq a \ln t + 0(a) + 0(t) + 0\left(\frac{st\ \ln(x + a + b)}{\ln t}\right) \\ &= a \ln a - a \ln(s + 1) + 0\left(\frac{a\ \ln(x + a + b)}{\ln a - \ln(s + 1)}\right)\end{aligned}$$

by (8.7.1), (8.7.2), and (8.7.3); since these involve only effective constants, the result follows.

□

Exercises

The following exercises are based on Gelfond [37], pp. 167-169, and lead to a proof that α and e^{α} cannot be simultaneously algebraic if $\alpha \neq 0$. They illustrate how some knowledge of the distribution of the primes can be helpful in obtaining transcendency results. The so-called Newton interpolation series that arise here are of historical significance in that they were used to obtain the first partial results on Hilbert's Seventh Problem (see Gelfond [39], pp. 102 and 103). An alternative proof of the "α, e^{α}" theorem will be found in Schneider [103], pp. 43-47 (see also Exercise 10.11).

<u>8.7.1</u>. Let T_N denote the least common multiple of $\{1,2,3,...,N\}$. Show that there is an absolute constant $C > 0$ such

that $\ell n\ T_N < CN$. (Hint: Show that

$$T_N = \prod_{p \le N} p^{[(\ell n\ N)/\ell n\ p]},$$

where p denotes a prime.)

8.7.2. Fix an integer $q \ge 1$. Then for any nonnegative integer n we can write $n = mq + r$, where $0 \le r \le q - 1$. Let $z_n = m$. Thus the sequence $\{z_n\}$ consists of q zeros followed by q ones, etc. Set $\varphi(z) = e^{\alpha z}$. In the notation of Exercise 8.3.10 show that, for any $\alpha \in \mathbb{C}$ with $300|\alpha| < q$,

$$e^{\alpha z} = \sum_{n=0}^{\infty} [z_0, \ldots, z_n](z - z_0)\cdots(z - z_n)$$

in the sense of uniform convergence. (Hint: Fix a $\rho > 0$. It suffices to show in Exercise 8.3.10 that $|R_n(z)| \to 0$ as $n \to \infty$ provided $|z| \le \rho$. Let $r = 3z_n$; then $r > 10\rho$ for n sufficiently large. Now

$$|R_n(z)| = \frac{1}{2\pi} \left| \int_{|\zeta|=r} \prod_{i=0}^{n} \frac{z - z_i}{\zeta - z_i} \frac{\exp(\alpha\zeta)}{\zeta - z} d\zeta \right|$$

$$\le \frac{r}{r - \rho} \exp(|\alpha| r) \prod_{i=0}^{n} \frac{\rho + z_i}{r - z_i}$$

$$\le 2\left\{\frac{\rho + z_n}{r - z_n}\right\}^{n+1} \exp\left(3|\alpha|\frac{n}{q}\right) \le 2\left\{\frac{1}{2}(3\rho r^{-1} + 1)\right.$$

$$\left. \cdot \exp\left(\frac{1}{100}\right)\right\}^{n+1} \to 0.)$$

*8.7.3. Show that the conclusion of Exercise 8.7.2 holds for $|\alpha| < q\ \ell n\ 2$. Is this best possible? (Hint: See Gelfond [37], pp. 148-154.)

<u>8.7.4.</u> Show that α and e^{α} cannot be simultaneously algebraic unless $\alpha = 0$. (Hint: Use Exercise 8.7.2. Set $A_n = [z_0,\dots,z_n]$ and write $n = mq + r$. Then, by Exercise 8.3.5,

$$A_n = \frac{1}{2\pi i}\int_{|\zeta|=mq} \frac{\exp(\alpha\zeta)\,d\zeta}{[\zeta(\zeta - 1)\cdots(\zeta - (m - 1))]^q(\zeta - m)^r}$$

so

$$|A_{mq+r}| < mq\exp(|\alpha|mq)[(mq)\cdots(m(q - 1) + 1)]^{-q}(m(q - 1))^{-r}.$$

By Stirling's estimate and Exercise 8.7.1 we find that if

$$\gamma = q!(m!)^q T_m^{2(q-1)} A_{mq+r},$$

then $|\gamma| < \exp(-\frac{1}{2}qm\,\ell n\,q)$ provided q is large and m is much larger than q (make this assumption throughout). From the hint to Exercise 8.3.6,

$$A_{mq+r} = \sum_{k=0}^{m}\sum_{\ell=0}^{p_k-1} \frac{\alpha^{p_k-1-\ell} e^{\alpha k}}{(p_k - 1 - \ell)!}\frac{1}{2\pi i}\int_{\mathcal{C}_k} \frac{(\zeta - k)^{p_k-\ell-1}}{\prod_{i=0}^{m}(\zeta - i)^{p_k}}\,d\zeta$$

$$= \sum_{k=0}^{m}\sum_{\ell=0}^{p_k-1} B_r(k,\ell,m)\alpha^{p_k-1-\ell}\beta^k,$$

where $p_0 = \cdots = p_{m-1} = q$, $p_m = r$, and $\mathcal{C}_k$ is the contour $|\zeta - k| = \frac{1}{2}$. By the reasoning of Lemma 8.4.1 there is an absolute constant $K_1 > 0$ such that $|B_r(k,\ell,m)| < K_1^{mq}(m!)^{-q}$. Hence if we set

$$C_r(k,\ell,m) = q!(m!)^q T_m^{2(q-1)} B_r(k,\ell,m),$$

we have $|C_r(k,\ell,m)| < K_2^{mq}$ by Exercise 8.7.1, where K_2 is an absolute constant. Now $C_r(k,\ell,m)$ is an integer since

$$C_r(k,\ell,m) = \frac{q!}{\ell!(p_k - 1 - \ell)!}(m!)^q T_m^{2(q-1)} D^\ell[(\zeta - k)^{p_k} \cdot \prod_{i=0}^{m} (\zeta - i)^{-p_i}]|_{\zeta=k}$$

$$= \frac{q!}{\ell!(p_k - 1 - \ell)!} \Sigma' \frac{\ell!}{\ell_0!\cdots\ell_m!} \Pi' \frac{(p_i + \ell_i - 1)!}{(p_i - 1)!} \cdot \frac{(m!)^q}{\Pi'(k - i)^{p_i}} \frac{T_m^{2(q - 1)}}{\Pi'(k - i)^{\ell_i}} .$$

By Lemma 4.1.7 and

$$\gamma = \sum_{k=0}^{m} \sum_{\ell=0}^{p_k-1} C_r(k,\ell,m)\alpha^{p_k-1-\ell}\beta^k,$$

either $\gamma = 0$ or

$$|\gamma| > K_3^{-m-q}((m + 1)qK_2^{mq})^{-K_4} \geq e^{-K_5 mq},$$

where K_3, K_4, and K_5 are constants depending only on α and β. However, for q sufficiently large, the latter alternative contradicts the upper bound on γ. Hence $A_n = 0$ for n sufficiently large and $e^{\alpha z}$ is a polynomial, a contradiction.)

Note that the above proof does not make any use of the pigeonhole principle. Although the pigeonhole principle is presently indispensable for obtaining results of any generality in

Diophantine approximation, it is artificial in some sense. In special cases where it can be avoided by means of an explicit continued fraction expansion, an integral identity, or an identity between hypergeometric series, it is sometimes possible to obtain much stronger results. For examples of the latter two techniques see the papers of Mahler, Siegel, and Baker referred to in Section 5.2, and also Baker [6], Bundschuh [17], Mahler [65], [71], and Osgood [79]. Bundschuh uses a Kummer relation for ${}_1F_1(a;b;x)$ to show (inter alia) that there are positive constants $c_1 < c_2$ such that

$$\left|e - \frac{p}{q}\right| > c_1 \frac{(\log \log q)}{q^2 \log q}$$

for all rational p/q, while

$$\left|e - \frac{p}{q}\right| < c_2 \frac{(\log \log q)}{q^2 \log q}$$

has infinitely many solutions.

8.7.5. Does the preceding result indicate that e is a typical or atypical real number with regard to approximation by rational numbers? (Hint: See Exercise 7.5.9.)

Chapter 9

FELDMAN'S THEOREM

9.1. Statement of the Theorem

In his papers [33] and [34] Feldman sharpened an earlier result of A. Baker's to obtain the following theorem:

Theorem 9.1.1. Let $\alpha_1,\ldots,\alpha_n$ be algebraic numbers whose logarithms are linearly independent over $\underset{\sim}{Q}$. Let $\gamma_0,\gamma_1,\ldots,\gamma_m$ be algebraic numbers, not all zero, of degree at most n. Then there are effectively computable numbers $c_i = c_i(m,n,\alpha_1,\ldots,\alpha_m) > 0$ $(1 \le i \le 2)$ such that

$$(9.1.1) \qquad \Gamma = |\gamma_0 + \gamma_1 \log \alpha_1 + \cdots + \gamma_m \log \alpha_m| \ge c_1/H^{c_2},$$

where $H = \max_{i=0}^{m} H(\gamma_i)$.

Here log z denotes any fixed determination of the logarithm of z, and z^β denotes $\exp(\beta \log z)$; we shall use $H(\gamma)$ and $L(\gamma)$ to denote the height and length of γ, respectively. When $m = 1$ and $\gamma_1 = -1$, we see immediately from (9.1.1) that $\log \alpha_1$ is transcendental for $\alpha_1 \ne 0,1$.

Many interesting consequences of Theorem 9.1.1 will be given in Section 10, and in Section 12 we shall see that it actually leads to an algorithm for solving Diophantine equations of the form (1.3). Whether (9.1.1) is valid for some $c_2 = c_2(m,n)$

independent of the α_i seems to be an open question; it is not hard to show that $c_2 \geq m$ (see Exercise 9.1.1).

Theorem 9.1.1 can be stated more precisely as follows:

Theorem 9.1.2. Let $n_0 = d(\underset{\sim}{K})$, where $\underset{\sim}{K} = \underset{\sim}{Q}(\alpha_1,\ldots,\alpha_m, \gamma_0,\gamma_1,\ldots,\gamma_m)$ and

$$M_1 = \max\{L(\alpha_1),\ldots,L(\alpha_m),\exp(|\log \alpha_1|),\ldots,\exp(|\log \alpha_m|)\}.$$

Then

$$(9.1.2) \qquad \Gamma > M_1^{-c'n_0m^5} \exp(-M_1^{c'n_0m^5} \ell n\, M_0),$$

where $M_0 = \max_{i=0}^{m} L(\gamma_i)$ and c' is an effectively computable absolute constant.

Theorem 9.1.2 will be deduced from Theorem 9.1.3.

Definition 9.1.1. Let $\beta_0,\beta_1,\ldots,\beta_{m-1}$ be algebraic numbers with $\beta_i \neq 0$ for some i and $|\beta_i| \leq 1$ for $i = 1,\ldots,m-1$. Let $M = \max_{i=0}^{m-1} L(\beta_i)$ and $n = \deg \underset{\sim}{Q}(\alpha_1,\ldots,\alpha_m,\beta_0,\ldots,\beta_{m-1})$.

Theorem 9.1.3 (Feldman). Under the above hypotheses,

$$(9.1.3) \qquad |\beta_0 + \beta_1 \log \alpha_1 + \cdots + \beta_{m-1} \log \alpha_{m-1} - \log \alpha_m| > \exp(-\mathcal{L} \log M),$$

where $\mathcal{L} = \lambda^{5m+9}$ and λ is any number that satisfies

$$(9.1.4) \qquad \lambda \geq \exp\{\theta(m^3 \ell n(m + 1) + nm\, \ell n\, M_1 + m(\ell n\, m)(\ell n\, M_1))\},$$

$\theta \geq 1$ being an effectively computable absolute constant.

To shorten the formulas involved in the proof, we make the following definition:

Definition 9.1.2. Let $J_0 = [\lambda \ln M] + 1$, $I_0 = [\lambda^{2+m^{-1}}] + 1$, $E = [\lambda^2]$, $q = [3\lambda^3 \ln M] + 1$, and $P_k(z) = (z + k + 1)\cdots(z + k + q - 1)$. Here $[x]$ denotes the greatest integer in x.

The proof itself, roughly speaking, is based on the auxiliary function of Section 8.6, namely,

$$(9.1.5)\quad g(z) = \sum_{k=0}^{q-1} \sum_{e_m=0}^{E-1} \cdots \sum_{e_1=0}^{E-1} P_k(z)(\alpha_1^{e_1}\cdots\alpha_m^{e_m})^z C_{ke_1\cdots e_m},$$

and proceeds as in Sections 8.2 and 8.4. In it one assumes that

$$(9.1.6)\quad |\beta_0 + \beta_1 \log \alpha_1 + \cdots + \beta_{m-1} \log \alpha_{m-1} - \log \alpha_m| \le \exp(-\mathcal{L} \log M).$$

Actually, Feldman uses Baker's idea of studying the more general auxiliary function

$$(9.1.7)\quad g(z_0,z_1,\ldots,z_m) = \Sigma\cdots\Sigma\, P_k(z_0)\alpha_1^{e_1z_1}\cdots\alpha_m^{e_mz_m}\, C_{ke_1\cdots e_m}.$$

One chooses the integers $C_{ke_1\cdots e_m}$ to make as many derivatives of g as possible vanish (or very small). At first it seems that, to control all derivatives of order ≤ 1, we must satisfy κI^{m+1} linear equations for some $\kappa > 0$. However, and this is the decisive point, we really need to satisfy only κI^m equations, for (9.1.6) allows us to use an auxiliary function with <u>one less variable</u>. More precisely, if $\lambda = \lambda(m,n,M_1)$ is large,

$$\begin{aligned}\alpha_1^{e_1}\cdots\alpha_m^{e_m} &= \exp(e_1 \log \alpha_1 + \cdots + e_m \log \alpha_m)\\ &\cong \exp\{e_1 \log \alpha_1 + \cdots + e_{m-1} \log \alpha_{m-1}\\ &\qquad + e_m(\beta_0 + \beta_1 \log \alpha_1 + \cdots + \beta_{m-1} \log \alpha_{m-1})\}\\ &= e^{\beta_0 e_m}\alpha_1^{e_1+e_m\beta_1}\cdots\alpha_{m-1}^{e_{m-1}+e_m\beta_{m-1}},\end{aligned}$$

and we introduce yet another auxiliary function,

$$f(z_0,z_1,\ldots,z_{m-1}) = \Sigma\cdots\Sigma\, P_k(z_0)e^{\beta_0 e_m z_0}\alpha_1^{(e_1+e_m\beta_1)z_1}\cdots \alpha_{m-1}^{(e_{m-1}+e_m\beta_{m-1})z_{m-1}}C_{ke_1\cdots e_m}. \tag{9.1.8}$$

Functions (9.1.7) and (9.1.8) are approximately equal along the diagonal $z_0 = z_1 = \ldots = z_m = z$, and it is here that we study them and their derivatives.

<u>Definition 9.1.3</u>. Let

$$\begin{Bmatrix} f_{i_0,i_1,\ldots,i_{m-1}}(z) \\ \\ g_{i_0,i_1,\ldots,i_{m-1}}(z)\end{Bmatrix} = \tag{9.1.9}$$

$$\Sigma\cdots\Sigma \sum_{t=0}^{i_0} \binom{i_0}{t} P_k^{(t)}(z)(\beta_0 e_m)^{i_0-t}\prod_{\eta=1}^{m-1}\{(e_\eta + \beta_\eta e_m)^{i_\eta}\alpha_\eta^{e_\eta z}\}$$

$$\cdot\begin{Bmatrix} e^{\beta_0 e_m z}\prod_{\eta=1}^{m-1}\alpha_\eta^{\beta_\eta e_m z} \\ \\ \alpha_m^{e_m z}\end{Bmatrix}\cdot C_{ke_1\cdots e_m}.$$

Thus $f_{i_0,i_1,\ldots,i_{m-1}}(z)$ is essentially a derivative of order $i_0 + i_1 + \cdots + i_{m-1}$ of $f(z_0,\ldots,z_{m-1})$ evaluated on the diagonal, and it is an approximation of $g_{i_0,i_1,\ldots,i_{m-1}}(z)$. Also $g_{0,\ldots,0}(z) = g(z)$.

The next definition uses the notation of Lemma 8.7.2.

Definition 9.1.4. For $z \in \underset{\sim}{Z}$ and $I = i_0 + i_1 + \cdots + i_{m-1}$, set $\gamma(z;I) = \gamma(q - 1, q - 1, z;I)$ and

$$\varphi = \varphi_{i_\eta} = \varphi_{i_0,i_1,\ldots,i_{m-1}}(z) = \gamma(z;I)^{-1} g_{i_0,i_1,\ldots,i_{m-1}}(z) \tag{9.1.10}$$

$$= \gamma^{-1} g_{i_\eta}(z).$$

This function φ is the really fundamental auxiliary function for the proof of Theorem 9.1.3; note that it is algebraic when $z \in \underset{\sim}{Z}$.

Many times during the proof of Theorem 9.1.3 we shall need a hypothesis of the form "λ exceeds a specific function of n, m, and M_1." On a first reading one should not attempt to verify the details, but rather to convince oneself in a general way that the argument is valid when λ is sufficiently large compared to n, m, and M_1. All of these hypotheses on λ are implied by (9.1.4), but for some purposes (9.1.4) is ridiculously overcautious. If the estimates are made with more care, hypothesis (9.1.4) can be weakened, but we shall not go into this.

Exercises

9.1.1. Let $\|y\|$ denote the distance from y to the nearest integer. Set $y_i(\underline{x}) = \sum_{j=1}^{m} a_{ij}x_j$ for $1 \le i \le n$, where the a_{ij} are real numbers and $\underline{x} = (x_1,\ldots,x_m)$ is a real vector. Show that for every $\varepsilon > 0$ and $B > 0$ there is an $H = H(\varepsilon) > B$ such that the system of $n + m$ inequalities

$$\|y_i(\underline{x})\| \le H^{-(m-\varepsilon)/n} \qquad (1 \le i \le n)$$

$$|x_j| \le H \qquad (1 \le j \le m)$$

has an integral vector solution $\underline{x} \ne 0$. (Hint: Introduce new variables $x_{m+1},\ldots,x_{m+n}$ and apply Exercise 2.2.4 to the system

$$y_i'(\underline{x}) = \begin{cases} y_i - x_{m+i} & \text{for} \quad 1 \le i \le n \\ x_{i-n} & \text{for} \quad n+1 \le i \le n+m \end{cases}$$

with

$$b_i = \begin{cases} H^{-(m-\varepsilon)/n} & \text{for} \quad 1 \le i \le n \\ H & \text{for} \quad n+1 \le i \le n+m.) \end{cases}$$

9.1.2. Show that $c_2 \ge m$ in Theorem 9.1.1. (Hint: It suffices to show that, for $c_2 < m$, the inequality

$$|\mathrm{Re}(\gamma_0 + \gamma_1 \log \alpha_1 + \cdots + \gamma_m \log \alpha_m)| \ge c_1/H^{c_2}$$

leads to a contradiction when the γ_i are integers with $|\gamma_i| \le H$. Use Exercise 9.1.1 with $\gamma_1,\ldots,\gamma_m$ playing the role of $x_1,\ldots,x_m$, $n = 1$, and ε suitably small.)

*9.1.3. Under the hypothesis of Exercise 9.1.1, show that, for $H > 1$, the system of $n + m$ inequalities

$$\|y_i(\underline{x})\| < H^{-m/n} \qquad (1 \le i \le n)$$

$$|x_j| \le H \qquad (1 \le j \le m)$$

has an integral vector solution $\underline{x} \ne 0$. (Hint: This can be obtained from Minkowski's theorem on convex bodies; see Cassels [18], p. 13.)

9.1.4. Deduce from Theorem 9.1.1 that the logarithm of an algebraic number cannot be a Liouville number.

**9.1.5. Can the logarithm of an S-number be a Liouville number?

9.2. Application of the Pigeonhole Principle

We now choose the $C_{ke_1 \cdots e_m}$ to make φ very flat; then we can proceed as in Chapter 8. For x real,

$$\varphi(x) = \sum_{k=0}^{q-1} \sum_{e_m=0}^{E-1} \cdots \sum_{e_1=0}^{E-1} S(k, e_1, \ldots, e_m), \tag{9.2.1}$$

where

$$S(k, e_1, \ldots, e_m) = \sum_{t=0}^{i_0} \binom{i_0}{t} (\beta_0 e_m)^{i_0 - t} \prod_{\eta=1}^{m-1} (e_\eta + \beta_\eta e_m)^{i_\eta} \cdot P \cdot \Pi_\alpha \cdot C,$$

$$P = P_k^{(t)}(x)/\gamma(q - 1, q - 1, x; I),$$

$$\Pi_\alpha = \prod_{\eta=1}^{m} \alpha_\eta^{e_\eta x},$$

$$C = C_{ke_1 \cdots e_m},$$

$$I = i_0 + i_1 + \cdots + i_{m-1}.$$

In most of what will follow P, Π_α, and C will be the only "important" parts of φ; both the other factors and also the number of terms will be negligible.

Clearly $\varphi(x)$ is linear in the $C_{ke_1\cdots e_m}$. We shall choose them to make both the real and imaginary parts of (9.2.1) very small for <u>integral</u> x and I such that $0 \le x \le J_0 - 1$ and $0 \le I \le I_0 - 1$. There are qE^m variables and at most $2J_0I_0^m$ linear forms. Thus we want $qE^m > 2J_0I_0^m$. Also, both for simplicity and to make the denominator of P large, we shall take q much larger than E, J_0, and I_0. One way of doing this would be to take

$$(9.2.2) \quad J_0 = [\lambda], \quad I_0 = [\lambda^2], \quad E = [\lambda^2], \quad q = [\lambda^3],$$

where $\lambda = \lambda(m,n,M_1)$ is a large parameter. Moreover (and this is important) the above conditions are still satisfied if $I_0 = [\lambda^{2+m^{-1}}]$. The actual choices of J_0, I_0, E, and q were made in Definition 9.1.2; note that J_0 and q also go to infinity with M.

Now set

$$(9.2.3) \quad \varphi = \sum_{k=0}^{q-1} \sum_{e_m=0}^{E-1} \cdots \sum_{e_1=0}^{E-1} \ell(i_0,i_1,\ldots,i_{m-1};x;k,e_1,\ldots,e_m) \cdot C_{ke_1\cdots e_m}.$$

Then

$$(9.2.4)\quad L \equiv \sum_k \sum_{e_m} \cdots \sum_{e_1} |\ell(\cdots)| \leq qE^m 2^I (2E)^{Im} \max(1, |\beta_0|)^I$$

$$\cdot\ M_1^{mEx} \max_{0 \leq t \leq I} P_{q-1}^{(t)}(x)(q-1)^{-q+1}$$

$$\cdot\ \exp\{(q-1)(\ell n(I+1) + c \frac{\ell n(x + 2q - 2)}{\ell n((q-1)/(I+1))})\}$$

by Lemma 8.7.2. We shall estimate L for $0 \leq x \leq J_0 - 1$ and $0 \leq I \leq I_0 - 1$. Note that $m \geq 1$ and λ, $\mathcal{L}$, M, and M_1 are all at least 2. First,

$$E^{Im} \leq \exp(\lambda^{2+m^{-1}} \cdot 2m\ \ell n\ \lambda).$$

Next, by (9.1.6),

$$|\beta_0| \leq m\ \ell n\ M_1 + M^{-\mathcal{L}} \leq (m+1)\ell n\ M_1$$

since $1/M \leq 1/2 < \ell n\ 2 \leq \ell n\ M_1$, so

$$\max(1, |\beta_0|)^I \leq \max(1, \exp\{\lambda^{2+m^{-1}} (\ell n(m+1) + \ell n\ \ell n\ M_1)\}).$$

Next,

$$(9.2.5)\qquad M_1^{mEx} \leq \exp(m\lambda^2 \cdot \lambda\ \ell n\ M \cdot \ell n\ M_1)$$

and

$$\exp\{(q-1)\ell n(I+1)\} \leq \exp\{3\lambda^3\ \ell n\ M \cdot \ell n(2\lambda^{2+m^{-1}})\}.$$

For $0 \leq t \leq I$, we have the estimate

$$P_{q-1}^{(t)}(x)(q-1)^{-q+1} \le \frac{P_{q-1}^{(t)}(x)}{(q-1)!} \le \binom{q-1}{t} \frac{(x+q)\cdots(x+2q-2)}{(q-1)!}$$

$$\le q^t \frac{(x+2q-2)!}{(x+q-1)!(q-1)!} \le q^I 2^{x+2q-2}$$

$$\le \exp\{\lambda^{2+m^{-1}} \ln(6\lambda^3 \ln M)$$

$$+ (\ln 2)(\lambda \ln M + 6\lambda^3 \ln M)\}.$$

Finally, $(q-1)/(I+1) \ge 2 \ln M \ge 2 \ln 2 > 1$, so

$$\frac{\ln(x+2q-2)}{\ln((q-1)/(I+1))} \le \frac{\ln(7\lambda^3 \ln M)}{\ln(2 \ln M)} \le \theta_1 \ln \lambda,$$

where θ_1 is an absolute constant and

$$\exp\{(q-1)c \frac{\ln(x+2q-2)}{\ln((q-1)/(I+1))}\}$$

$$\le \exp(3c\lambda^3 \ln M\cdot\theta_1 \ln \lambda).$$

Thus (qE^m is negligible) there is an absolute constant $\theta_2 > 1$ such that

$$(9.2.6) \quad L \le \exp\{\theta_2\lambda^3 \ln M\cdot(m \ln M_1 + (2+m^{-1})\ln \lambda)\} = L_1.$$

Henceforth θ_k, where k is a positive integer, shall denote an effectively computable absolute constant.

Note: As further motivation for Definition 9.1.2 we see from (9.2.4) and (9.2.5) that insofar as an estimate of type (9.2.6) is concerned, EJ_0 might as well be as large as q.

Let $C \geq 1$ be an integer. By Lemma 2.2.2 (the pigeonhole principle) we can choose integers $C_{ke_1 \cdots e_m}$, not all zero, such that

$$|C_{ke_1 \cdots e_m}| \leq C \tag{9.2.7}$$

and

$$|\mathrm{Re}(\varphi)|, \ |\mathrm{Im}(\varphi)| \leq \frac{2L_1 C}{(C + 1)^r - 2} \tag{9.2.8}$$

when

$$0 \leq x \leq J_0 - 1 \quad \text{and} \quad 0 \leq I \leq I_0 - 1, \tag{9.2.9}$$

provided $r = qE^m/2J_0I_0^m > 1$. But

$$r > \frac{(3\lambda^3 \ \ell n \ M)(\lambda^2 - 1)^m}{(2\lambda \ \ell n \ M)\lambda^{2m+1}} = \frac{3}{2}\,\lambda(1 - \lambda^{-2})^m > \frac{1}{2}\,\lambda > 1,$$

provided

$$\lambda > (1 - 3^{-1/m})^{-1/2}. \tag{9.2.10}$$

<u>Note</u>: We would not have $r > 1$ if I_0^m were replaced by I_0^{m+1}. See comment after (9.1.7).

Thus, under the assumptions (9.2.9) and (9.2.10),

$$|\varphi| = |\varphi_{i_0, i_1, \ldots, i_{m-1}}(x)| \leq 4L_1 C\{(C + 1)^{\lambda/2} - 2\}^{-1}. \tag{9.2.11}$$

Now choose $C = [M^{\lambda^3}]$. Then if

$$\lambda > 8 + 2^2 \cdot 5^4 \theta_2^2 \ m \ \ell n \ M_1, \tag{9.2.12}$$

it follows that

$$|\varphi| \leq \exp(-\theta_3\lambda^4 \ \ell n \ M) \tag{9.2.13}$$

for integral x and I in the range (9.2.9).

Exercises

The following trivial exercises are typical of the large number of very simple estimations used in the proof of Theorem 9.1.3:

9.2.1. If $k_1, k_2 \geq 1$ and $\lambda > 2(k_1 + k_2)(1 + \ell n \ k_2)$, show that

$$\lambda > k_1 + k_2 \ \ell n \ \lambda .$$

9.2.2. If $a > b > 0$, show that $(\ell n \ ax)/(\ell n \ bx)$ is decreasing.

9.2.3. If $x,y > 0$, show that

$$x > \ell n(1 + x) > y(1 - (1 + x)^{-1/y})$$

and hence $\ell n(2 \ \ell n \ 2) > \ell n(4 - 2\sqrt{2}) > 0.1$.

9.2.4. Show that, for $\lambda \geq 2$ and $x \geq 2 \ \ell n \ 2$,

$$(\ell n(3.5 \ \lambda^3 x))/(\ell n \ x) < 90 \ \ell n \ \lambda .$$

9.3. The Discreteness of φ

By Lemma 4.1.7 we shall show for certain integers j that if $|\varphi(j)|$ is nonzero, it cannot be too small. We first introduce the following definition:

<u>Definition 9.3.1</u>. Let

$$\Omega(A,B) = \{(j,i_\eta) = (j,i_0,i_1,\ldots,i_{m-1}) \mid j \text{ and the } i_\eta \text{ are integers, } 0 \le j \le A-1, \text{ and } 0 \le i_0 + i_1 + \cdots + i_{m-1} \le B-1\}.$$

For example, (9.2.13) holds for $(j,i_\eta) \in \Omega(J_0,I_0)$.

For $(j,i_\eta) \in \Omega(J_0,I_0)$, we see from (9.2.1) that $\alpha = \varphi_{i_0,i_1,\ldots,i_{m-1}}(j)$ is of the form (4.1.7) with $r = 2m$,

$$\alpha_\eta = \begin{cases} \alpha_\eta & \text{for } 1 \le \eta \le m, \\ \beta_{\eta-m-1} & \text{for } m+1 \le \eta \le 2m, \end{cases}$$

$$N_\eta \le \begin{cases} (E-1)(J_0-1) & \text{for } 1 \le \eta \le m, \\ i_0 & \text{for } \eta = m+1, \\ i_{\eta-m-1} & \text{for } m+2 \le \eta \le 2m, \end{cases}$$

and $1 \le n_\eta$ for $1 \le \eta \le 2m$. As a bound for the left-hand side of (4.1.8) we can take

$$B = \exp(\theta_3\lambda^3 \ln \lambda \ln M), \tag{9.3.1}$$

provided

$$\lambda > 9m + 1. \tag{9.3.2}$$

The estimations proceed almost exactly as in Section 9.2, except that $\beta_0,\beta_1,\ldots,\beta_{m-1}$ and Π are not involved; note that

$$|C_{ke_1\cdots e_m}| \le \exp(\lambda^3 \ln M)$$

and that the number of terms is negligible because

$$\prod_{\eta=1}^{2m} (N_\eta + 1) \leq 2^{2m} I_0^m (E - 1)^m (J_0 - 1)^m.$$

Thus, for $\alpha \neq 0$, we have

$$|\alpha| \geq \exp\{-(n - 1)\theta_3\lambda^3 \ \ell n\ \lambda\ \ell n\ M$$

$$- nm\lambda^3\ \ell n\ M\ \ell n\ M_1 - 2nm\lambda^{2+m^{-1}}\ \ell n\ M\}$$

since $1 \leq n_\eta$, $L_2(\alpha_\eta) \leq L(\alpha_\eta)$,

$$M_1^{-nm(E-1)(J_0-1)} \leq \prod_{\eta=1}^{m} L^{-nN_\eta}(\alpha_\eta),$$

and

$$M^{-nmI_0} \leq \prod_{\eta=1}^{m} L^{-nN_{\eta+m}}(\beta_{\eta-1}).$$

It follows that

$$(9.3.3) \qquad |\alpha| = |\varphi_{i_0,i_1,\ldots,i_{m-1}}(j)| \geq \exp(-\theta_4\lambda^3\ \ell n^2\ \lambda\ \ell n\ M),$$

provided that $\alpha \neq 0$, $(j,i_\eta) \in \Omega(J_0,I_0)$, and

$$(9.3.4) \qquad \lambda > \exp\{(10 + 2\theta_3)nm\ \ell n\ M_1\}.$$

9.4. The Flatness of φ

For

$$(9.4.1) \qquad \lambda > \max\{e^{12},(\theta_4/\theta_3)^{12}\}$$

the lower bound of (9.3.3) exceeds the upper bound of (9.2.13). Hence we have the following lemma:

Lemma 9.4.1. There are $C_{ke_1\cdots e_m} \in \underset{\sim}{Z}$, not all zero, such that

(9.4.2) $$|C_{ke_1\cdots e_m}| \le [M^{\lambda^3}]$$

and

(9.4.3) $$\varphi_{i_0,i_1,\ldots,i_{m-1}}(j) = 0 \quad \text{for} \quad (j,i_\eta) \in \Omega(J_0,I_0).$$

Thus φ is very flat along an initial segment of the x-axis.

We shall now show that $\varphi = \gamma(x,I)^{-1}g$ is approximated very closely by $\gamma(x,I)^{-1}f$; in other words,

(9.4.4) $$d(x) = |f_{i_0,i_1,\ldots,i_{m-1}}(x) - g_{i_0,i_1,\ldots,i_{m-1}}(x)|$$

is small if x is not too large. In fact, when z is a complex variable and $I = i_0 + i_1 + \cdots + i_{m-1} \le 2\lambda^3$, we have

$$d(z) = |\Sigma\cdots\Sigma \sum_{t=0}^{i_0} \binom{i_0}{t} P_k^{(t)}(z)(\beta_0 e_m)^{i_0-t} \prod_{\eta=1}^{m-1}$$

$$\cdot\{(e_\eta + \beta_\eta e_m)^{i_\eta} \alpha_\eta^{e_\eta z}\} C_{ke_1\cdots e_m}$$

$$\cdot\ (\exp\{e_m z(\beta_0 + \beta_1 \log \alpha_1 + \cdots + \beta_{m-1} \log \alpha_{m-1})\}$$

$$- \exp\{e_m z(\log \alpha_m)\})|$$

$$\le qE^m 2^I \max_{0\le t\le I} P_{q-1}^{(t)}(|z|) \max(1,(m+1)\ln M_1)^I (2E)^{Im}.$$

$$\cdot \exp(mE|z| \ln M_1 + \lambda^3 \ln M)$$

$$\cdot |\exp\{e_m z(\beta_0 + \beta_1 \log \alpha_1 + \cdots + \beta_{m-1}$$

$$\cdot \log \alpha_{m-1} - \log \alpha_m)\} - 1|.$$

Now $|e^z - 1| \le |z|e^{|z|}$ (use the infinite series expansion of e^z), so by (9.1.6) the last factor above is at most

$$E|z| \exp(-\mathcal{L} \ln M) \exp\{E|z| \exp(-\mathcal{L} \ln M)\}.$$

Finally, as in Section 9.2, we obtain

$$\max_{0 \le t \le I} P_{q-1}^{(t)}(|z|) \le (q - 1)! q^{I_2|z|+2q-1}$$

and (since $\lambda > 2(m + 1)\ln M_1$ by (9.3.4)) also

$$\text{(9.4.5)} \quad d(z) \le (q - 1)! \exp\{\theta_5(m\lambda^3 \ln \lambda \ln M + |z|(1 + \lambda^2 m \ln M_1)) - \mathcal{L} \ln M\} \equiv B(\theta_5, |z|).$$

This will be very useful when $\mathcal{L}$ is large in comparison to $\lambda^3 \ln \lambda$ and $\lambda^2|z|$.

9.5. Application of the Hermite Interpolation Formula

We are now ready to use the extrapolation technique of Chapter 8.

Definition 9.5.1. Let $J_u = [\lambda^{1+(u/3m)} \ln M] + 1$ and $I_u = [2^{-u}\lambda^{2+m^{-1}}] + 1$, where $0 \le u \le U$ and

$$U = 13m^2 + 12m.$$

This agrees with Definition 9.1.2 for $u = 0$. Note that, as functions of u, J_u is increasing and I_u is decreasing. The following lemma (essentially due to A. Baker) is the key to the proof of Feldman's theorem.

<u>Lemma 9.5.1</u>. If

$$(9.5.1) \qquad \varphi_{i_0,i_1,\ldots,i_{m-1}}(j) = 0 \quad \text{for} \quad (j,i_\eta) \in \Omega(J_u,I_u),$$

where $0 \leq u \leq U - 1$, then (9.5.1) also holds for $(j,i_\eta) \in \Omega(J_{u+1},I_{u+1})$. It also follows that

$$(9.5.2) \qquad \max_{|z|\leq J_{u+1}} |g_{i_0,i_1,\ldots,i_{m-1}}(z)|$$

$$\leq (q-1)! \exp(-2^{-u}\lambda^{3+\{(u+3)/3m\}} \ln M)$$

for $I \leq I_{u+1} - 1$.

<u>Proof</u>. Let i,i_η,j denote nonnegative integers. We begin by estimating the derivatives of

$$f(z) = f_{i_\eta}(z) = f_{i_0,\ldots,i_{m-1}}(z)$$

$$= (\log \alpha_1)^{-i_1} \cdots (\log \alpha_{m-1})^{-i_{m-1}} \Sigma \cdots \Sigma$$

$$\cdot [P_k(z)e^{\beta_0 e_m z}]^{(i_0)}$$

$$\cdot [\alpha_1^{(e_1+\beta_m e_m)z}]^{(i_1)}$$

$$\cdot \ldots [\alpha_{m-1}^{(e_{m-1}+\beta_{m-1}e_m)z}]^{(i_{m-1})} C_{ke_1\cdots e_m}.$$

Suppose that $0 \leq i_0 + \cdots + i_{m-1} \leq I_{u+1} - 1$ and $0 \leq i \leq I_{u+1} - 1$. Then by Leibniz's rule (see Exercise 9.5.2),

$$(9.5.3)\quad f^{(i)}(z) = \Sigma \frac{i!}{h_0!\cdots h_{m-1}!} (\log \alpha_1)^{-i_1} \cdots (\log \alpha_{m-1})^{-i_{m-1}} \cdot \Sigma\cdots\Sigma(\cdots)^{(i_0+h_0)} \cdots (\cdots)^{(i_{m-1}+h_{m-1})} C_{ke_1\cdots e_m}$$

$$= \Sigma \frac{i!}{h_{\eta}!} (\log \alpha_1)^{h_1} \cdots (\log \alpha_{m-1})^{h_{m-1}} f_{i_\eta + h_\eta}(z),$$

where $h_\eta! = h_0!\cdots h_{m-1}!$ and the first sum is over all nonnegative-integer m-tuples $(h_0,\ldots,h_{m-1})$ such that $h_0 + h_1 + \cdots + h_{m-1} = i$. Now

$$(i_0 + h_0) + \cdots + (i_{m-1} + h_{m-1}) \leq 2(I_{u+1} - 1) \leq I_u - 1,$$

so, for integers j with $0 \leq j \leq J_{j-1}$, we have

$$(9.5.4)\quad |f_{i_\eta+h_\eta}(j)| = |f_{i_\eta+h_\eta}(j) - g_{i_\eta+h_\eta}(j)|$$
$$\leq (q - 1)! \exp\{\theta_5(m\lambda^3 \ln \lambda \ln M \cdot + J_u(1 + \lambda^2 m \ln M_1)) - \mathfrak{L} \ln M\}$$
$$= B(\theta_5, J_u)$$

by (9.1.10), (9.5.1), and (9.4.5). Now assume that

$$(9.5.5)\qquad \lambda > (e + \ln M_1)^m.$$

We deduce from (9.5.4) that

$$|f^{(i)}(j)| \leq B(\theta_8, J_u) \quad \text{for} \quad 0 \leq j \leq J_u - 1,\ 0 \leq i \leq I_{u+1} - 1.$$

Simply note that the number of terms in the last line of (9.5.3) is at most $(i + 1)^m$, the multinomial coefficients are bounded by m^i (see Exercise 9.5.2), and

$$\begin{aligned}(i + 1)^m m^i \max_\eta |\log \alpha_\eta|^{mi} &\\ &\le \exp(\theta_6 \lambda^3 \ln \lambda + 2m\lambda^3 \ln \ln M_1) \\ &\le \exp(\theta_7 \lambda^3 \ln \lambda).\end{aligned}$$

Next one can obtain

$$|f^{(i)}_{i_0,\ldots,i_{m-1}}(z) - g^{(i)}_{i_0,\ldots,i_{m-1}}(z)| \le B(\theta_9, |z|)$$

by the same sort of argument as in Section 9.4. It follows then that

$$(9.5.6) \quad |g^{(i)}_{i_0,\ldots,i_{m-1}}(j)| \le B(\theta_{10}, J_u) \quad \text{for} \quad 0 \le j \le J_u - 1, \quad 0 \le i \le I_{u+1} - 1.$$

We now apply the Hermite formula to show (recall Definition 9.1.4) that

$$(9.5.7) \quad \gamma(j,I)\varphi_{i_\eta}(j) = g_{i_\eta}(j) = g_{i_0,\ldots,i_{m-1}}(j) = 0$$

for $J_u \le j \le J_{u+1} - 1$ and $0 \le I \le I_{u+1} - 1$. By Theorem 8.3.1,

(9.5.8) $$g_{i_\eta}(z) = \sum_{j=0}^{J_u-1} \sum_{i=0}^{I_{u+1}-1} \frac{g_{i_\eta}^{(i)}(j)}{i!} \frac{1}{2\pi i} \int_{|\zeta-j|=1/2} (\zeta - j)^i (z - \zeta)^{-1} \cdot \prod_{\nu=0}^{J_u-1} \left(\frac{z - \nu}{\zeta - \nu}\right)^{I_{u+1}} d\zeta$$

$$+ \frac{1}{2\pi i} \int_{|\zeta|=J_{u+2}} g_{i_\eta}(\zeta)(\zeta - z)^{-1} \cdot \prod_{\nu=0}^{J_u-1} \left(\frac{z - \nu}{\zeta - \nu}\right)^{I_{u+1}} d\zeta$$

$$\equiv \mathcal{M} + \mathcal{S},$$

where $\mathcal{M}$ is the "main term" and $\mathcal{S}$ is the integral on $|\zeta| = J_{u+2}$. Let z be a complex number such that $J_u \leq |z| \leq J_{u+1}$.

The number of terms in $\mathcal{M}$ is

$$J_u I_{u+1} \leq 2\lambda^3 \cdot 2\lambda^{1+(u/3m)} \ln M,$$

and this is negligible. By Corollary 8.4.1,

(9.5.9) $$\left| \prod_{\nu=0}^{J_u-1} \left(\frac{z - \nu}{\zeta - \nu}\right)^{I_{u+1}} \right| \leq (8 \cdot 2^{2J_{u+1}+5J_u})^{I_{u+1}}$$

$$\leq \exp(\theta_{12} 2^{-u} \lambda^{3+\{(u+4)/3m\}} \ln M).$$

Also $|(z - \zeta)^{-1}| \leq 2$, so (9.5.6) yields

$$(9.5.10)\quad |\mathcal{M}| \le J_u I_{u+1} B(\theta_{10}, J_u)$$

$$\cdot \tfrac{1}{2}\cdot 2\cdot \exp\{\theta_{12} 2^{-u} \lambda^{3+\{(u+4)/3m\}} \ln M\}$$

$$\le (q-1)! \exp\{\theta_{11} m\lambda^3 \ln\lambda \ln M$$

$$+ 2\theta_{11}\cdot 2\lambda^{1+\frac{u}{3m}} \ln M\cdot \lambda^2 m \ln M_1 - \mathcal{L} \ln M$$

$$+ \theta_{13} 2^{-u} \lambda^{3+\{(u+4)/3m\}} \ln M\}$$

$$\le (q-1)!\ \exp\{\theta_{14} 2^{-u} \lambda^{3+\{(u+4)/3m\}} \ln M - \mathcal{L} \ln M\},$$

provided

$$(9.5.11)\quad \lambda > \exp\{8m(\ln(1+2m))(1+\dot{U}+\ln(m \ln M_1))\}.$$

Next, the trivial estimate for $\mathcal{E}$ (the "error term") yields

$$(9.5.12)\quad |\mathcal{E}| \le \max_{|\zeta|\le J_{u+2}} |g_{i_\eta}(\zeta)|$$

$$\cdot \frac{J_{u+2}}{J_{u+2}-J_{u+1}} \left(\frac{J_{u+1}+J_u}{J_{u+2}-J_u}\right)^{J_u I_{u+1}}.$$

The same estimates that gave (9.4.5) yield

$$(q-1)!\ \exp\{\theta_5(m\lambda^3 \ln\lambda \ln M$$

$$+ 2\lambda^{1+\{(u+2)/3m\}} \ln M(1+\lambda^2 m \ln M_1))\}$$

as an upper bound for the first factor of (9.5.12) while the second factor is bounded by 2 if

$$(9.5.13)\qquad \lambda > 2^{12m}.$$

It also follows from (9.5.13) that the third factor is bounded by

$$\left(\theta_{15}\frac{J_{u+1}}{J_{u+2}}\right)^{J_u I_{u+1}} \le (\theta_{16}\lambda^{-1/3m})^{J_u I_{u+1}}$$

$$\le \exp\{-\theta_{17}m^{-1}2^{-u}\lambda^{3+\{(u+3)/3m\}} \ell n\, \lambda\, \ell n\, M$$

$$+ \theta_{18}2^{-u}\lambda^{3+\{(u+3)/3m\}} \ell n\, M\}.$$

Thus, for

$$\lambda > 2^{m+U}, \tag{9.5.14}$$

we have

$$|\mathcal{S}| \le (q - 1)!\ \exp\{\theta_5(m\lambda^3\ \ell n\, \lambda\ \ell n\, M$$
$$+ 2\lambda^{1+\{(u+2)/3m\}}\ \ell n\, M(1 + \lambda^2 m\ \ell n\, M_1))$$
$$- \theta_{19}m^{-1}2^{-u}\lambda^{3+\{(u+3)/3m\}}\ \ell n\, \lambda\ \ell n\, M\}. \tag{9.5.15}$$

<u>It is now clear that we really needed $I_0 \ge \lambda^{2+m^{-1}}$ rather than merely $I_0 > \lambda^2$.</u> Next, for

$$\lambda > (m^2 2^{1+U}\ \ell n\, M_1)^{3m}, \tag{9.5.16}$$

we have

$$|\mathcal{S}| \le (q - 1)!\ \exp\{-\theta_{20}m^{-1}2^{-u}\lambda^{3+\{(u+3)/3m\}}\ \ell n\, \lambda\ \ell n\, M\}. \tag{9.5.17}$$

Estimate (9.5.2) of the lemma now follows from (9.5.10), (9.5.17),

$$\mathcal{L} > \theta_{21}\lambda^{3+\{(u+4)/3m\}} \tag{9.5.18}$$

[see (9.1.3) and Definition 9.5.1], and

(9.5.19) $$\lambda > \exp\{6m(1 + \ln 3m)(2 + |\ln \theta_{20}|) + U + 2m/\theta_{20}\}.$$

Now, by applying Lemma 4.1.7 as in Section 9.3, we can show that if $\varphi_{i_\eta}(j) \neq 0$ for some integer j with $J_u \le j \le J_{u+1} - 1$, then

(9.5.20) $$\begin{aligned}|\varphi_{i_\eta}(j)| &\ge \exp\{-(n-1)\theta_3\lambda^3 \ln \lambda \ln M - (n-1)j \ln 2 \\ &\qquad - 2nm\lambda^2 j \ln M_1 - 2nm\lambda^{2+m^{-1}} \ln M\} \\ &\ge \exp\{-\theta_{22}\lambda^{3+\{(u+1)/3m\}} \ln \lambda \ln M\},\end{aligned}$$

provided

(9.5.21) $$\lambda > M_1^{3nm}.$$

On the other hand, $\varphi_{i_\eta}(j) = \gamma(j,I)^{-1} g_{i_\eta}(j)$, so by (8.7.5), Definition 9.1.4, (9.5.2), and Definition 9.5.1,

(9.5.22) $$\begin{aligned}|\varphi_{i_\eta}(j)| &\le (q-1)!\, \exp\{-2^{-u}\lambda^{3+\{(u+3)/3m\}} \ln M\} \\ &\qquad \cdot (q-1)!^{-1} \exp\{(q-1)(\ln(I+1) \\ &\qquad\qquad + c\, \frac{\ln(j+2q-2)}{\ln((q-1)/(I+1))})\} \\ &\le \exp\{-2^{-u}\lambda^{3+\{(u+3)/3m\}} \ln M + \theta_{23}\lambda^3 \ln \lambda \ln M \\ &\qquad + \theta_{24} \max\{(1 + \frac{u+1}{3m}), 3\}\, \lambda^3 \ln \lambda \ln M\} \\ &\le \exp\{-\theta_{25}\lambda^{3+\{(u+3)/3m\}} \ln M\},\end{aligned}$$

provided

(9.5.23) $\lambda > \exp\{10m(1 + \ln m)(1 + U + \ln(1 + \theta_{23} + \theta_{24}))\}$.

But for

(9.5.24) $\lambda > \exp\{3m(1 + \ln 2m)(1 + \ln(1 + \theta_{22}\theta_{25}^{-1}))\}$

the bounds (9.5.20) and (9.5.22) are inconsistent. Hence (9.5.7) holds, and together with the hypothesis of the lemma this shows that

$$\varphi_{i_\eta}(j) = 0 \quad \text{for} \quad (j, i_\eta) \in \Omega(J_{u+1}, I_{u+1}).$$

□

Note: Lemma 8.7.2 really played an important role, for in (9.5.22) it allowed us to get rid of the $(q - 1)!$ that came from (9.5.2). Had the $(q - 1)!$ remained, there would have been an

$$\exp(\lambda^3 \ln M \ln \ln M)$$

factor in (9.5.22), and this would have ruined the argument.

Exercises

9.5.1 (Multinomial theorem). Show that

$$\left(\sum_{j=1}^{m} x_j\right)^i = \sum \frac{i!}{i_1! \cdots i_m!} x_1^{i_1} \cdots x_m^{i_m},$$

where the sum on the right is over all nonnegative integers $i_1, \ldots, i_m$ such that $i_1 + \cdots + i_m = i$.

9.5.2. The expressions $i!/i_1! \cdots i_m!$ occurring on the right-hand side of the formula of Exercise 9.5.1 are called multinomial coefficients of order i. Show that they are bounded by m^i. (Hint: Set $x_j = 1$ for $1 \le j \le m$.)

9.5.3 (Leibniz's rule). If $f_j(x)$ has n derivatives for $1 \le j \le m$, show that

$$\frac{\partial^n}{\partial x^n} \prod_{j=1}^{m} f_j(x) = \sum \frac{i!}{i_1! \cdots i_m!} \frac{\partial^{i_1} f_1}{\partial x^{i_1}} \cdots \frac{\partial^{i_m} f_m}{\partial x^{i_m}} ;$$

here the summation convention is the same as in Exercise 9.5.1.

9.6. Conclusion of the Proof

The differentiated functions of Definition 9.1.3 have served their purpose, and we return to the function $g(z) = g_{0,\ldots,0}(z)$ given by (9.1.5). By Lemma 9.5.1,

$$g(j) = 0 \qquad (j \in \underset{\sim}{Z},\ 0 \le j \le J_U - 1), \tag{9.6.1}$$

where

$$g(z) = \sum_{k=0}^{q-1} \sum_{e_m=0}^{E-1} \cdots \sum_{e_1=0}^{E-1} P_k(z)(\alpha_1^{e_1} \cdots \alpha_m^{e_m})^z C_{ke_1 \cdots e_m} .$$

Now (9.6.1) is a set of $J_U \ge \lambda^{1+(U/3m)} \ln M$ linear equations in $qE^m \le 6\lambda^{2m+3} \ln M$ unknowns $C_{ke_1 \cdots e_m}$. So for

$$\lambda^{1+(U/3m)} \ge 6\lambda^{2m+3},$$

in particular when

$$U \ge 3m(2m + 4), \tag{9.6.2}$$

the determinant of the system of J' equations

$$g(j) = 0 \qquad (0 \le j \le J' - 1),$$

must vanish for some $J' \le J_U$. From Lemma 8.5.2 and Section 8.6, it follows that

$$(9.6.3) \quad \alpha_1^{e_1} \cdots \alpha_m^{e_m} = \alpha_1^{e'_1} \cdots \alpha_m^{e'_m} \quad \text{for} \quad (e_1, \ldots, e_m) \ne (e'_1, \ldots, e'_m),$$

where $0 \le e_i, e'_i \le E - 1$. Hence

$$(9.6.4) \qquad \sum_{i=1}^{m} (e_i - e'_i) \log \alpha_i = 2\pi i \nu \qquad (\nu \in \underset{\sim}{Z}).$$

This would be an immediate contradiction if we knew that $\log \alpha_1, \ldots, \log \alpha_m, \log(-1)$ were linearly independent. However, this assumption is not necessary. We shall replace α_i by $\alpha_i^{2^{-r}}$ $(1 \le i \le m)$, where

$$r = [\log_2(4\lambda^2 m \ \ell n\ M_1)] \le \theta_{26}\ \ell n\ \lambda$$

and show that the left-hand side of (9.6.4) is so small that $\nu = 0$. This will contradict the independence of the log α_i.

Let

$$h(z) = \frac{g(2^{-r}z) 2^{(r+1)(q-1)}}{(q-1)!}$$

We shall show the following:

Lemma 9.6.1. For integers j such that

$$(9.6.5) \qquad 0 \le j \le 6\lambda^{2m+3}\ \ell n\ M + \lambda,$$

we have

$$h(j) = 0. \tag{9.6.6}$$

Proof. By (9.5.2),

$$|h(j)| \le \exp\{\theta_{27}\lambda^3 \ln \lambda \ln M - 2^{-U}\lambda^{3+\{(U+3)/3m\}} \ln M\} \tag{9.6.7}$$

since $6\lambda^{2m+3} \ln M + \lambda \le J_U$ by (9.6.2). On the other hand, we can calculate a good lower bound for h(j) if it is nonzero. By Lemma 8.7.1,

$$u = \frac{P_k(j2^{-r})2^{(r+1)(q-1)}}{(q-1)!}$$

$$= \{\frac{2^{q-1}}{(q-1)!}\} \prod_{i=1}^{q-1} (j + k2^r + i2^r) \in \underset{\sim}{Z}$$

and by (9.6.5)

$$|u| \le 2^{(r+1)(q-1)} \frac{([j2^{-r} + 2q - 2] + 1)!}{[j2^{-r} + q - 1]!(q-1)!} \tag{9.6.8}$$

$$\le 2^{(r+1)(q-1)}2^{j2^{-r}+2q-1}$$

$$\cdot \frac{([j2^{-r} + 2q - 2] - [j2^{-r} + q - 1] + 1)!}{(q-1)!}$$

$$\le q(q+1) \exp\{(\ln 2)((r+1)(q-1)$$

$$+ j2^{-r} + 2q - 1)\}$$

$$\le \exp(\theta_{28}\lambda^3 \ln \lambda \ln M + \theta_{29}\lambda^{2m+1} \ln M).$$

We now apply Lemma 4.1.7 to h(j) with the "r" of that lemma set equal to m, the α_i replaced by $\alpha_i^{2^{-r}}$, $N_i \le \lambda^{2m+5} \ell n\ M + \lambda^3$, $1 \le n_i$, and n replaced by

$$n' = d[\underset{\sim}{Q}(\alpha_1^{2^{-r}},\ldots,\alpha_m^{2^{-r}})/\underset{\sim}{Q}] \le (2^r n)^m.$$

As a bound for the left-hand side of (4.1.8) we can take

$$(9.6.9) \quad B = \exp\{\theta_{30}(\lambda^3\ \ell n\ \lambda\ \ell n\ M + \lambda^{2m+1}\ \ell n\ M)\}$$

by (9.4.2) and (9.6.8). Since by Lemma 4.1.5(iv),

$$\ell n\ L(\alpha_i^{2^{-r}}) \le \ell n(2^{n2^r} L(\alpha_i)) \le n2^r + \ell n\ M_1,$$

it follows that

$$(9.6.10) \quad |h(j)| \ge B^{1-n'} \prod_{i=1}^{m} L^{-n'N_i}(\alpha_i^{2^{-r}})$$

$$\ge B^{1-n'} \exp\{-(n2^r)^m(\lambda^{2m+5}\ \ell n\ M + \lambda^3) \cdot (n2^r + \ell n\ M_1)\}$$

$$\ge \exp\{-\theta_{30}(n' - 1) \cdot (\lambda^3\ \ell n\ \lambda\ \ell n\ M + \lambda^{2m+1}\ \ell n\ M) - n^m(4m\ \ell n\ M_1)^{m+1}(\lambda^{4m+7}\ \ell n\ M + \lambda^{2m+5}) \cdot (n + \ell n\ M_1)\}.$$

Now for

$$(9.6.11) \qquad U = 13m^2 + 12m$$

and

$$(9.6.12) \quad \lambda > 2^{90m} + (1 + \theta_{27} + \theta_{30})^2 (2mn \ \ell n \ M_1)^{12},$$

the bounds (9.6.7) and (9.6.10) are inconsistent, so $h(j) = 0$.

□

Just as we obtained (9.6.4) from (9.6.1), we now obtain

$$(9.6.13) \qquad 2^{-r} \sum_{i=1}^{m} (e_i - e_i') \log \alpha_i = 2\pi i \nu \qquad (\nu \in \underset{\sim}{Z}),$$

from (9.6.6) (take $t = 2^{-r}$ in Lemma 8.5.2). But the left-hand side of (9.6.13) is less than

$$\frac{2}{4\lambda^2 m \ \ell n \ M_1} \cdot m \cdot 2\lambda^2 \ \ell n \ M_1 = 1 < 2\pi,$$

so $\nu = 0$. This contradicts the linear independence of the $\log \alpha_i$. Hence (9.1.6) is untenable, and Theorem (9.1.3) follows.

Next we prove Theorem (9.1.2). Consider

$$\Gamma = |\gamma_0 + \gamma_1 \log \alpha_1 + \cdots + \gamma_m \log \alpha_m|$$

as in (9.1.1). Without loss of generality we may assume that $|\gamma_m| = \max_{i=1}^{m} |\gamma_i|$. Set $\beta_i = \gamma_i \gamma_m^{-1}$ for $i = 0,\ldots,m$. Then by Theorem (9.1.3),

$$\Gamma \geq |\gamma_m| \exp(-\mathcal{L} \ \ell n \ M).$$

Now

$$M = \max L(\beta_i) = \max L(\gamma_i \gamma_m^{-1}) \leq 2^{n^2} M_0^{2n}$$

by Lemma 4.1.5(ii). If $|\gamma_m| \geq 1$, the result follows immediately. Otherwise $\gamma = \gamma_m$ is the root of its minimal polynomial,

$$a_n z^n + a_{n-1} z^{n-1} + \cdots + a_0,$$

so $|a_0| = |a_n\gamma^n + \cdots + a_1\gamma| \leq \gamma(M_0 - |a_0|)$ and $\gamma \geq 1/M_o$. Hence

$$\Gamma \geq M_0^{-1} \exp\{-\mathcal{L}\ \ell n(2^{n^2} M_0^{2n})\}$$

$$\geq c_1(m,n,M_1) M_0^{-c_2(m,n,M_1)}$$

with $c_2 = \exp\{\theta_{31}(m^2\ \ell n(m+1) + nm^2\ \ell n\ M_1 + m^2(\ell n\ m)(\ell n\ M_1))\}$ and $c_1 = c_2^{-1}$, the constant θ_{31} being an absolute constant. Since $n_0 \geq n$, the result follows.

□

Chapter 10

TRANSCENDENCE THEOREMS OF BAKER

Theorem 9.1.1 now enables us to prove the transcendence of various classes of numbers. For this purpose we only need to know that Γ in (9.1.1) cannot vanish. As before, log z will denote any fixed determination of the logarithm of z.

Theorem 10.1. If α_i is algebraic for $1 \leq i \leq m$ and $\log \alpha_1, \ldots, \log \alpha_m$ are linearly independent over $\underset{\sim}{Q}$, then

$$1, \log \alpha_1, \ldots, \log \alpha_m$$

are linearly independent over $\underset{\sim}{A}$.

Proof. By Theorem (9.1.1). □

Theorem 10.2. If α_i and β_i are algebraic for $1 \leq i \leq m$ and the α_i are nonzero, then

$$\beta_1 \log \alpha_1 + \cdots + \beta_m \log \alpha_m$$

is either zero or transcendental.

Proof. Say

$$\beta_0 + \beta_1 \log \alpha_1 + \cdots + \beta_m \log \alpha_m = 0 \qquad (\beta_0 \in \underset{\sim}{A}). \tag{10.1}$$

By Theorem 10.1, the logarithms are dependent over $\underset{\sim}{Q}$, so we can eliminate them successively. It is clear that at each step the coefficients of the $\log \alpha_i$ will always remain algebraic. Hence the left-hand side of (10.1) equals β_0.

□

Theorem 10.3. If $\alpha_1,\ldots,\alpha_m;\beta_0,\beta_1,\ldots,\beta_m$ are nonzero algebraic numbers, then

$$e^{\beta_0}\alpha_1^{\beta_1}\cdots\alpha_m^{\beta_m}$$

is transcendental.

Proof. Say

$$e^{\beta_0}\alpha_1^{\beta_1}\cdots\alpha_m^{\beta_m} = \alpha \in \underset{\sim}{A}.$$

Then

$$\beta_0 + \beta_1 \log \alpha_1 + \cdots + \beta_m \log \alpha_m - \log \alpha$$
$$- \beta \log(-1) = 0 \qquad (\beta \in \underset{\sim}{Z}).$$

Hence by Theorem 10.2 we have $\beta_0 = 0$, a contradiction.

□

Theorem 10.4. If, for $1 \le i \le m$, the α_i and β_i are algebraic, $\alpha_i \ne 0$ or 1 for all i, and $1,\beta_1,\ldots,\beta_m$ are linearly independent over $\underset{\sim}{Q}$, then

$$\alpha_1^{\beta_1}\cdots\alpha_m^{\beta_m}$$

is transcendental.

Proof. It suffices to show that

$$\zeta = \zeta_1^{\beta_1}\cdots\zeta_m^{\beta_m}$$

is transcendental for some $N \in \mathbb{Z}^+$, where $\zeta_i^N = \alpha_i$, $1 \le i \le m$. For N sufficiently large,

$$\beta_1 \operatorname{Log} \zeta_1 + \cdots + \beta_m \operatorname{Log} \zeta_m + (-1) \operatorname{Log} \zeta = 0, \tag{10.2}$$

where Log z denotes the principal value of the logarithm of z (see Exercise 10.1). We now introduce a certain reduction process. Say

$$\rho_1 \operatorname{Log} \sigma_1 + \cdots + \rho_j \operatorname{Log} \sigma_j = 0,$$

where $\sigma_i, \rho_i \in \mathbb{A}$ and the ρ_i are linearly independent over $\mathbb{Q}$. Then by Theorem 10.1 there are rational numbers r_i, not all zero, $1 \le i \le j$, such that

$$r_1 \operatorname{Log} \sigma_1 + \cdots + r_j \operatorname{Log} \sigma_j = 0.$$

Without loss of generality, $r_j \neq 0$, so

$$(\rho_1 - r_1 r_j^{-1} \rho_j) \log \sigma_1 + \cdots$$
$$+ (\rho_{j-1} - r_{j-1} r_j^{-1} \rho_j) \log \sigma_{j-1} = 0,$$

and the coefficients of the log σ_i are again linearly independent over $\mathbb{Q}$. Now assume that ζ is algebraic. By applying the reduction process to (10.2) we find that Log $\zeta = 0$ or Log $\zeta_i = 0$ for some i, $1 \le i \le m$. Hence Log $\zeta = 0$. But on applying it once again, we do obtain Log $\zeta_i = 0$ for some i, and this is a contradiction. □

<u>Corollary 10.1</u>. The number log α is transcendental if $\alpha \neq 1$ is algebraic.

<u>Proof</u>. By Theorem 10.2. □

<u>Corollary 10.2 (Hilbert's Seventh Problem)</u>. The number α^{β} is transcendental if α is algebraic, $\alpha \neq 0,1$ and β is irrational.

<u>Proof</u>. By Theorem 10.4. □

For example, $2^{\sqrt{2}}$ is transcendental.

<u>Corollary 10.3</u>. The numbers e, π, and e^{π} are transcendental.

<u>Proof</u>. By Theorem 10.3 ($\alpha_i = \beta_i = 1$ for all i), Corollary 10.1 ($\alpha = -1$), and Corollary 10.2 ($e^{\pi} = e^{-2i(i\pi/2)} = i^{-2i}$). □

<u>Corollary 10.4</u>. The number $\exp(\alpha\pi + \beta)$ is transcendental if $\alpha,\beta \in \underset{\sim}{A}$ and $\beta \neq 0$.

<u>Proof</u>. By Theorem 10.3 ($\exp(\alpha\pi + \beta) = e^{\beta}e^{-2i\alpha(i\pi/2)}$ $= e^{\beta}i^{-2i\alpha}$). □

<u>Corollary 10.5</u>. The number $\pi + \log \alpha$ is transcendental if α is algebraic.

<u>Proof</u>. By Theorem 10.2 and Corollary 10.3 (if $\pi + \log \alpha = 0$, then $e^{-\pi} = \alpha$, which is a contradiction). □

All results obtained in this chapter are special cases of the following:

<u>Schanuel's Conjecture</u>. If $\zeta_1,\ldots,\zeta_m \in \underset{\sim}{C}$ are linearly independent over $\underset{\sim}{Q}$, then

$$\underset{\sim}{Q}(\zeta_1,\ldots,\zeta_n,e^{\zeta_1},\ldots,e^{\zeta_n})$$

has transcendence degree at least n.

For example, the <u>algebraic</u> independence of π and e is an easy consequence of this. Ax [1] has actually proved the function field analog of this conjecture.

The results of this section occur in the fundamental papers of Baker [5] on which the work of Feldman in Chapter 9 is based. Baker's work in turn may be considered as the continuation of a program begun by Gelfond ([39], p. 177) with his solution of Hilbert's Seventh Problem. For the history of this problem, which was also solved independently by T. Schneider, see Gelfond [39], pp. 102-109.

Exercises

10.1. Prove (10.2). (Hint: Choose N so that

$$|\beta|(\ell n|\alpha| + \pi)/N < \pi/2m,$$

where $|\alpha| = \max_{i=1}^{m} |\alpha_i|$ and $|\beta| = \max_{i=1}^{m} |\beta|$.)

10.2. Deduce Theorem 10.1 from Schanuel's conjecture.

10.3. Show that $I_n = 1 - \frac{1}{n+1} + \frac{1}{2n+1} - \frac{1}{3n+1} + \cdots$ is transcendental for $1 \leq n \leq 3$. (Hint:

$$I_n = \int_0^1 \frac{1}{1+x^n}\,dx \quad \text{and} \quad f'(x) = \frac{1}{1+x^3}$$

if

$$f(x) = \frac{1}{6}\,[2\log(1+x) - \log(1-x+x^2) + 2\sqrt{3}\tan^{-1}\frac{2x-1}{\sqrt{3}}].$$

For n = 3, this is listed by Siegel [110], p. 97 as an unsolved problem. In fact I_n is transcendental for $n \in \underset{\sim}{Z}^+$; see Exercise 10.4.)

10.4 (van der Poorten [85]). Let P(z) and Q(z) denote polynomials with algebraic coefficients and with no common polynomial factor. Denote by $\alpha_1,\ldots,\alpha_n$ the distinct roots of Q(z) and by $\gamma_1,\ldots,\gamma_n$ the residues at the poles $\alpha_1,\ldots,\alpha_n$ of the rational function P(z)/Q(z). Furthermore, let $\mathcal{C}$ be a contour for which

$$\int_{\mathcal{C}} = \int_{\mathcal{C}} \frac{P(z)}{Q(z)}\, dz$$

exists and suppose that $\mathcal{C}$ is either closed or has endpoints that are algebraic or infinite. Show that $\int_{\mathcal{C}}$ is algebraic if and only if

$$\int_{\mathcal{C}} \sum_{m=1}^{n} \frac{\gamma_m}{z - \alpha_m}\, dz = 0.$$

As a consequence, show that if $d(P) < d(Q)$ and the roots of Q(z) are distinct, then

$$\int_{\mathcal{C}} \frac{P(z)}{Q(z)}\, dz$$

(provided it exists) is transcendental or zero.

10.5. Show that either $e + \pi$ or $e\pi$ is transcendental.

**10.6. Is π^e transcendental? What about $\ell n\, \ell n\, 2$ and $\ell n^2\, 2 + \ell n^2\, 3$?

The following exercises are based on Gelfond and Linnik [40], pp. 232-239; they yield "elementary" proofs of some of the results of this section.

10.7. Let a_0,a_1,b_0,b_1,b_2 be real numbers with $b_2 \neq 0$. Show that

$$f(x) = e^x(a_1x + a_0) + e^{2x}(b_2x^2 + b_1x + b_0)$$

can have at most four real zeros, counting multiplicities. (Hint: If it had five real zeros, then

$$\frac{d^2}{dx^2} e^{-x}f(x) = e^{2x}(b_2'x^2 + b_1'x + b_0')$$

would have three real zeros by Rolle's theorem.)

<u>10.8</u>. Let $P_0(x),\ldots,P_{q-1}(x) \in \underset{\sim}{R}[x]$ with $n_i = d(P_i)$ for $0 \leq i \leq q - 1$. Let $\gamma_0,\ldots,\gamma_{q-1}$ be distinct real numbers. Show

$$f(x) = \sum_{\ell=0}^{q-1} e^{\gamma_\ell x} P_\ell(x)$$

has at most $n_0 + \cdots + n_{q-1} + q - 1$ real zeros, counting multiplicities. (Hint: Say it has $n_0 + \cdots + n_{q-1} + q$ real zeros. Let $Df(x) = f'(x)$. Consider

$$g(x) = D^{n_{q-2}+1} \exp[(-\gamma_{q-2} + \cdots + (-1)^{q-1}\gamma_0)x]$$

$$\cdots D^{n_2+1} \exp[(-\gamma_2 + \gamma_1 - \gamma_0)x]D^{n_1+1}$$

$$\cdot \exp[(-\gamma_1 + \gamma_0)x]D^{n_0+1} \exp[-\gamma_0 x]f(x).$$

This is a polynomial of degree n_{q-1} multiplied by an exponential. But by Rolle's theorem it has $n_{q-1} + 1$ zeros.

<u>10.9</u>. Let $f(x)$ be a real-valued function on $[a,b]$ such that its first I derivatives exist and are continuous. Show that if $f(x)$ has at least I zeros in $[a,b]$, then

$$|f^{(i)}(x)| \leq \frac{(b-a)^{I-i}}{(I-i)!} \max_{a \leq x \leq b} |f^{(I)}(x)|.$$

(Hint: First prove this in the case where f(x) has a single zero in [a,b] of multiplicity N and then generalize.)

10.10. In the notation of Exercise 10.8 let $n = \max(n_i + 1)$ and assume that (i) $q \leq n^2$, (ii) $H(P_\ell) \leq \exp(K_1 n^2)$ for $0 \leq \ell \leq q-1$, and (iii) $|\gamma_\ell| \leq K_2 n$ for $0 \leq \ell \leq q-1$, where K_1 and K_2 are independent of n. Say f(x) has at least I zeros in $[0,n^{1/2}]$, where $I \geq K_3 n^2$ and K_3 is independent of n and I. Show that for n sufficiently large

$$|f^{(i)}(x)| < n^{-\frac{1}{5}I}$$

provided $0 \leq i \leq I/\ln n$ and $0 \leq x \leq n^{1/2}$. (Hint: Use Exercise 10.9 and Stirling's estimate:

$$f^{(i)}(x) = \sum_{\ell=0}^{q-1} \sum_{\nu=0}^{i} \binom{i}{\nu} \gamma_\ell^{\nu} e^{\gamma_\ell x} P_\ell^{(i-\nu)}(x)$$

for all $i \geq 0$, so by Exercise 10.9

$$|f^{(i)}(x)| \leq \frac{n^{\frac{1}{2}I}}{(0.9I)!} q(I+1)2^I(K_2 n)^I e^{K_2 n^{3/2}} \cdot n! \cdot n e^{K_1 n^2} n^{\frac{1}{2}n}$$

$$< n^{-\frac{1}{5}I} \quad \text{for n sufficiently large.)}$$

10.11. Show that if α is a real algebraic number and $\alpha \neq 0$, then e^{α} is transcendental. (Hint: We can assume that α and $e^{\alpha} = \beta$ are algebraic integers of degree and height d, H and d', H' respectively. Let

$$f(x) = \sum_{k=0}^{q-1} \sum_{\ell=0}^{q-1} C_{k\ell} x^k e^{\alpha \ell x},$$

where the integers $C_{k\ell}$ will be chosen later. By Leibniz's rule,

$$f^{(i)}(x) = \sum_{k=0}^{q-1} \sum_{\ell=0}^{q-1} C_{k\ell} \sum_{\nu=0}^{i} \binom{i}{\nu} \frac{k!}{(k-\nu)!} x^{k-\nu} (\alpha\ell)^{i-\nu} e^{\alpha \ell x}.$$

Next, if N is a nonnegative integer,

$$\alpha^N = \sum_{s=0}^{d-1} a_s \alpha^s,$$

where $a_s = a_s(N) \in \underset{\sim}{Z}$ and $|a_s| \leq (H+1)^N$ by Lemma 4.1.4; an analogous result holds for β. Thus, if x is a nonnegative integer,

$$(*) \quad f^{(i)}(x) = \sum_{s=0}^{d-1} \sum_{t=0}^{d'-1} \{ \sum_{k=0}^{q-1} \sum_{\ell=0}^{q-1} C_{k\ell} D(i,x,k,\ell,s,t) \} \alpha^s \beta^t,$$

where $D = D(i,x,k,\ell,s,t) \in \underset{\sim}{Z}$ and

$$|D| \leq (i+1) 2^i i! 2^k x^k q^i dd'(H+1)^i (H'+1)^{qx}.$$

Henceforth K_i shall denote a quantity independent of q. If $i \leq q^2/\ell n\, q$ and $0 \leq x \leq \ell n^2 q$, then

$$|D| \leq \exp(K_4 q^2).$$

By Lemma 2.2.1 we can choose integers $C_{k\ell}$ with $|C_{k\ell}| \leq \exp(K_5 q^2)$ such that the $\{\ \}$ term in (*) vanishes for

$$0 \leq i \leq q^2/\ell n\ q \text{ and } 0 \leq x \leq (\ell n\ q)/2dd'.$$

But now if q is large, f(x) satisfies the hypotheses of Exercise 10.10 (in Exercise 10.10, q is to be replaced by q^2, n by q, and I by $q^2/2dd'$), so

$$|f^{(i)}(x)| < \exp(-\frac{1}{10dd'} q^2\ \ell n\ q)$$

for

$$(**) \qquad 0 \leq i \leq q^2/2dd'\ \ell n\ q \quad \text{and} \quad 0 \leq x \leq \ell n^2\ q < q^{1/2}.$$

On the other hand, Lemma 4.1.7 implies that either $f^{(i)}(x) = 0$ or

$$|f^{(i)}(x)| \geq \exp(-K_6 q^2)$$

when (**) holds. Hence for q large we must have $f^{(i)}(x) = 0$ when (**) holds. Hence the number of real roots of f(x) grows like $(q^2\ \ell n\ q)/2dd'$. But by Exercise 10.8, f(x) has at most $q^2 - 1$ real roots, a contradiction.)

10.12. Show that if α and β are positive real algebraic numbers and neither is 1, then $\theta = (\ell n\ \alpha)/(\ell n\ \beta)$ is either rational or transcendental. (Hint: Proceed as in Exercise 10.11, with

$$f(x) = \sum_{k=0}^{q-1} \sum_{\ell=0}^{q-1} C_{k\ell}\alpha^{kx}\beta^{\ell x}.$$

Note that in Exercise 10.8 the γ_i must be distinct.)

10.13. Show that $\sin \alpha$ and $\cos \alpha$ are both transcendental when α is a nonzero algebraic number. (Hint: $\sin^2 \alpha + \cos^2 \alpha = 1$.)

10.14. For what algebraic α is $\tan \alpha$ algebraic? (Hint: First determine when β and $\tan^{-1} \beta$ are simultaneously algebraic, and express $\tan^{-1}$ in terms of logarithms.)

10.15. Show that the following general statements are equivalent:

(1) If $\alpha_1, \ldots, \alpha_n \in \underset{\sim}{A}$ are distinct, then $e^{\alpha_1}, \ldots, e^{\alpha_n}$ are linearly independent over $\underset{\sim}{A}$.

(2) If $\alpha_1, \ldots, \alpha_n \in \underset{\sim}{A}$ are linearly independent over $\underset{\sim}{Q}$, then $e^{\alpha_1}, \ldots, e^{\alpha_n}$ are algebraically independent.

*10.16. Show that the statements of Exercise 10.15 are true. (Hint: This is Lindemann's theorem; a nice exposition can be found in Niven [78], pp. 117-133.)

10.17. If t is a real number, let $t^{1/3}$ denote the real cube root of t. Let $\alpha = (2 + \sqrt{5})^{1/3}$ and $\beta = (2 - \sqrt{5})^{1/3}$. Are α and β linearly independent over $\underset{\sim}{Q}$? Are α, β, and 1 linearly independent over $\underset{\sim}{Q}$?

10.18. Let $f(t) = \Sigma_{i=1}^{k} a_i e^{\alpha_i t}$, where the a_i are not all zero and the α_i are not all zero. Show that among the 4k - 1 numbers $f(j)$ and $f^{(i)}(0)$, where i and j are integers satisfying $0 \leq i, j \leq 2k - 1$, at least one is transcendental.

Chapter 11

EFFECTIVE ALGEBRAIC NUMBER THEORY

11.1. Definitions and Preliminary Results

Since we are interested in algorithms that actually solve certain Diophantine equations (see Chapter 12), we shall develop the rudiments of algebraic number theory in a constructive way. Much of mathematics can be treated this way (see Bishop [11] and Péter [83], pp. 269-281). Our own treatment owes much to Borevich and Shafarevich [13], pp. 75-129.

As in Definition 3.2.1, we say that we can effectively do something if we have an algorithm for doing it.

Definition 11.1.1. The set $S \subseteq \mathbb{C}$ is an effectively approximable set if for every $s \in S$ and $N \in \mathbb{Z}^+$ we can effectively find an $r \in \mathbb{Q}(i)$ such that

$$|s - r| < 1/N. \tag{11.1.1}$$

Definition 11.1.2. The set $S \subseteq \mathbb{C}$ is an effectively bounded below set if (i) for every $s \in S$ we can effectively decide whether or not $s = 0$ and (ii) we can effectively find an $N \in \mathbb{Z}^+$ such that

$$|s| > 1/N \tag{11.1.2}$$

for all $s \in S$ with $s \neq 0$.

Note: Let $s = .a_1a_2a_3\ldots$ and $\pi = 3.b_1b_2b_3\ldots$ in decimal notation, where

$$a_i = \begin{cases} 1 & \text{if} \quad b_{i+10^{10}} = 7, \\ 0 & \text{otherwise.} \end{cases}$$

Then $S = \{s\}$ is effectively approximable, but it is an open question whether or not S is effectively bounded below because it is unknown if the digit 7 occurs in the decimal expansion of π beyond the first 10 million places.

In the following we say that s is effectively approximable if $S = \{s\}$ is effectively approximable, etc.

It is clear that s is effectively approximable if and only if its real and imaginary parts are effectively approximable, and that if s is effectively approximable and real, we may take $b = 0$ in (11.1.1). It is also easy to show (see Exercise 11.1.1) that if s_1 and s_2 are effectively approximable, then so are $-s_1$, $s_1 + s_2$, and s_1s_2.

Lemma 11.1.1. If $s \neq 0$ is effectively approximable and effectively bounded below, then s^{-1} is effectively approximable.

Proof. We first show this for real $s > 0$. Since s is effectively bounded below, there is an $N \in \underset{\sim}{Z}^+$ such that $s \geq 1/N$. Since s is effectively approximable, we can find, for $M \in \underset{\sim}{Z}^+$, some rational r such that

$$|s - r| < 1/N(2 + 2MN) \equiv h. \tag{11.1.3}$$

[This implies that $(1 + 2MN)h < r$.] Now

$$(r + h)^{-1} < s^{-1} < (r - h)^{-1},$$

and from

$$(r - h)^{-1} - (r + h)^{-1} = \frac{2h}{(r - h)(r + h)}$$

$$< \frac{2h}{(2MNh)(2 + 2MN)h} < \frac{1}{M}$$

it follows that s^{-1} is effectively approximable. Next, if $s = x + iy$ is complex, it is clear that $|s|^2 = x^2 + y^2$ is effectively approximable and effectively bounded below; hence $(x^2 + y^2)^{-1}$ is effectively approximable, and so is

$$s^{-1} = (x - iy)(x^2 + y^2)^{-1}.$$

□

The next lemma shows that, given a finite set of positive real numbers, we can "almost" find the smallest.

<u>Lemma 11.1.2</u>. Let S be a finite, effectively approximable, and effectively bounded below set of positive real numbers. Then, given $m \in \underset{\sim}{Z}^+$, we can effectively find an $s_1 \in S$ such that

(11.1.4) $$s_1/s < 1 + 1/m \quad \text{for} \quad \text{all } s \in S.$$

<u>Proof</u>. Choose $N \in \underset{\sim}{Z}^+$ so that $s \geq 1/N$ for all $s \in S$. Let $S = \{s_1, \ldots, s_n\}$ and choose $r_i \in \underset{\sim}{Q}$ with $r_i \geq 0$ so that

$$|s_i - r_i| < 1/(3 + 2m)N \equiv h.$$

Without loss of generality we may assume that $0 \leq r_1 \leq \cdots \leq r_n$. Then

$$(11.1.5) \qquad s_1/s_i < (r_1 + h)(r_i - h)^{-1} \leq (r_1 + h)(r_2 - h)^{-1}.$$

Now $r_2 \geq r_1 > (2m + 1)h$ implies that

$$(11.1.6) \qquad m(r_1 + h) < (m + 1)(r_2 - h),$$

and the result follows from (11.1.5) and (11.1.6).

□

Lemma 11.1.3. Given $P(z) \in \underset{\sim}{Z}[z]$, we can effectively find irreducible polynomials $P_1(z),\ldots,P_r(z)$ with $r \leq n = d(P)$ such that

$$P(z) = P_1(z)\cdots P_r(z).$$

Proof. If P(z) is not irreducible, then

$$(11.1.7) \qquad P(z) = Q_1(z)Q_2(z),$$

where $d(Q_1),d(Q_2) \geq 1$ and $Q_1(z),Q_2(z) \in \underset{\sim}{Z}[z]$, by Theorem 4.1.1. Now $d(Q_1) \leq n - 1$ and $H(Q_1) \leq H(Q_1)H(Q_2) \leq c(n)H(P)$ by Theorem 4.1.2. Hence $Q_1(z)$ belongs to a finite set of polynomials that we can construct, and the same holds for $Q_2(z)$. Thus by testing all possibilities we can effectively find a factorization of the form (11.1.7). We repeat this process until P(z) is completely factored.

□

Lemma 11.1.4. If $\alpha_1,\ldots,\alpha_m$ are the distinct roots of the polynomial $P(z) \in \underset{\sim}{Z}[z]$, where $d(P) = n \geq m$, then we can effectively find an $N \in \underset{\sim}{Z}^+$ such that

(11.1.8) $$|\alpha_i - \alpha_j| > 1/N \qquad (1 \leq i < j \leq m).$$

Proof. As in the proof of Lemma 11.1.3, we can show that $L_2(\alpha_i) \leq c_1(n)L_2(P)$. Choose $N \in \underset{\sim}{Z}^+$ so that

$$N > n^2 3^{n^2} c_1(n)^{2n} L_2^{2n}(P).$$

By Lemma 4.1.5(iii) we have

$$d(\alpha_i - \alpha_j)L(\alpha_i - \alpha_j) \leq N,$$

and hence (11.1.8) follows by Lemma 3.2.1. □

Lemma 11.1.5. If $\alpha_1,\ldots,\alpha_m$ are the distinct roots of the polynomial $P(z) = a_n z^n + \cdots + a_0 \in \underset{\sim}{Z}[z]$ and $M \in \underset{\sim}{Z}^+$, then we can effectively find $r_j \in \underset{\sim}{Q}(i)$ such that

(11.1.9) $$|\alpha_j - r_j| < 1/M \qquad (1 \leq j \leq m \leq n).$$

Proof. Find an $N \in \underset{\sim}{Z}^+$ so that (11.1.8) holds. If $|z - \alpha_j| \geq 1/4MN$ for all j $(1 \leq j \leq m)$, then

(11.1.10) $$|P(z)| = |a_n \prod_{j=1}^{m} (z - \alpha_j)^{e_j}| \geq (\frac{1}{4MN})^n = A,$$

where e_j is the multiplicity of the root α_j $(1 \leq j \leq m)$. Let $x = \mathrm{Re}(z)$ and $y = \mathrm{Im}(z)$. The roots of $P(z)$ all lie in the rectangle $|x| \leq nH(P)$, $|y| \leq nH(P)$ by the argument of Lemma 3.2.1. Introduce the grids

$$G_c = \{(a + ib)/c \mid a,b \in \underset{\sim}{Z},\ c \in \underset{\sim}{Z}^+, \text{ and}$$

$$|a|, |b| \leq cnH(P)\} = \{r_k\}.$$

For each α_j $(1 \leq j \leq m)$, there is clearly an $r_j \in G_c$ such that $|\alpha_j - r_j| < 1/c$. Moreover,

$$(11.1.11) \qquad |P(r_j)| < |a_n| c^{-1} \prod_{k \neq j} |r_j - \alpha_j + \alpha_j - \alpha_k|^{e_k}$$

$$\leq c^{-1} H(P)[1 + 2nH(P)]^{n-1} = B.$$

We can effectively find a number $c \in \underset{\sim}{Z}^+$ so large that $B < A$. List all elements of the set

$$G'_c = \{r_k \mid r_k \in G_c \text{ and } |P(r_k)| < B\}.$$

Each $r \in G'_c$ satisfies

$$(11.1.12) \qquad |r - \alpha_j| < 1/4MN < 1/M$$

for some j $(1 \leq j \leq m)$; also, for each j $(1 \leq j \leq m)$, there is an $r \in G'_c$ such that (11.1.12) holds. Whenever (11.1.12) holds we say that r is near α_j. An r can be near only one α_i, since r near α_j, α_k, and $j \neq k$ implies

$$|\alpha_k - \alpha_j| \leq |\alpha_k - r| + |r - \alpha_j| < 1/2MN,$$

a contradiction by (11.1.8). We now effectively partition G'_c into equivalence classes. Say $r_1 \sim r_2$ if $|r_1 - r_2| < 1/2MN$. If r_1 and r_2 are near α_k and α_j, respectively, $k \neq j$, and $r_1 \sim r_2$, then

$$|\alpha_k - \alpha_j| \le |\alpha_k - r_1| + |r_1 - r_2| + |r_2 - \alpha_j|$$

$$\le \frac{\frac{1}{4} + \frac{1}{2} + \frac{1}{4}}{MN} \le \frac{1}{N},$$

a contradiction by (11.1.8). On the other hand, if r_1 and r_2 are both near α_i, then

$$|r_1 - r_2| \le |r_1 - \alpha_i| + |\alpha_i - r_2| < 1/2MN.$$

Hence $r_1 \sim r_2$ if and only if they are both near the same α_i. Finally, choose an r_j from each equivalence class; by (11.1.12) these satisfy (11.1.9).

□

Definition 11.1.3. We say that $\alpha \in \underset{\sim}{A}$ is effectively given if (i) its minimal polynomial P is given and (ii) an $r \in \underset{\sim}{Q}(i)$ is given so that

$$|\alpha - r| < 1/4N, \tag{11.1.13}$$

where $N \in \underset{\sim}{Z}^+$ satisfies (11.1.8) for P.

It is clear that if $\alpha,\beta \in \underset{\sim}{A}$ are effectively given, then we can determine whether or not $\alpha = \beta$.

Lemma 11.1.6. If $\alpha \in \underset{\sim}{A}$ is effectively given, it is effectively approximable and effectively bounded below.

Proof. Let $P(z) = a_n(z - \alpha_1)\cdots(z - \alpha_n)$ be the minimal polynomial of $\alpha = \alpha_1$. That α is effectively bounded below is immediate (Lemma 3.2.1). To show that it is effectively approximable, choose $r_j \in \underset{\sim}{Q}(i)$ as in (11.1.9) with $M > 4N$, where N satisfies (11.1.8). Let $r \in \underset{\sim}{Q}(i)$ satisfy (11.1.13). Then

$$|r_1 - r| \leq |r_1 - \alpha| + |\alpha - r| < 1/2N$$

whereas, for $j \geq 2$,

$$|r_j - r| = |(\alpha_j - \alpha) + (r_j - \alpha_j) + (\alpha - r)|$$
$$\geq 1/N - 1/4N - 1/4N = 1/2N.$$

Hence we can determine which r_j is r_1. □

Lemma 11.1.7. If $\alpha \in \underset{\sim}{A}$ is effectively approximable and we are given a polynomial $P(z) \in \underset{\sim}{Z}[z]$ such that $P(\alpha) = 0$, then α is effectively given.

Proof. By Lemma 11.1.3 we can effectively write

$$P(z) = aP_1^{e_1}(z)\cdots P_r^{e_r}(z) \qquad (a \in \underset{\sim}{Z}),$$

where $P_1(z),\ldots,P_r(z) \in \underset{\sim}{Z}[z]$ are relatively prime and irreducible, and the $e_i \geq 0$ are integers. Choose N as in (11.1.8) for P(z). Let $\alpha_1,\ldots,\alpha_{m_1}$ be the roots of $P_1(z)$ and choose $r_1,\ldots,r_{m_1} \in \underset{\sim}{Q}(i)$ so that

$$|\alpha_j - r_j| < 1/4N \qquad (1 \leq j \leq m_1);$$

this can be done effectively by Lemma 11.1.5. Choose $r \in \underset{\sim}{Q}(i)$ so that $|\alpha - r| < 1/4N$. As in the proof of Lemma 11.1.6, we can determine whether or not α is one of the α_j, where $1 \leq j \leq m_1$. Hence we can determine which $P_s(z)$ has α as a root. Condition (ii) of Definition 11.1.3 follows from the fact that α is effectively approximable. □

<u>Lemma 11.1.8</u>. If α, β are effectively given, so are $\bar{\alpha}$, $\alpha \pm \beta$, $\alpha\beta$, and α/β (if $\beta \neq 0$).

<u>Proof</u>. By Lemma 11.1.6, α and β are effectively approximable and effectively bounded below, so $\bar{\alpha}$, $\alpha \pm \beta$, $\alpha\beta$, and α/β are effectively approximable (see Lemma 11.1.1 for α/β). By means of the symmetric function theorem we can effectively produce polynomials in $Z[z]$ having $\bar{\alpha}$, $\alpha \pm \beta$, $\alpha\beta$, and α/β as roots. Hence the result follows from Lemma 11.1.7. □

<u>Lemma 11.1.9</u>. If $\alpha \in \underset{\sim}{A}$ is effectively given and $|\alpha| \neq 0,1$, then $\ell n|\alpha|$ is effectively approximable and effectively bounded below.

<u>Proof</u>. The numbers $\alpha\bar{\alpha}$ and $\alpha\bar{\alpha} - 1$ are effectively approximable and effectively bounded below by Lemmas 11.1.6 and 11.1.8. Thus we can effectively find an $N \in \underset{\sim}{Z}^+$ such that

$$\alpha\bar{\alpha}, \quad |\alpha\bar{\alpha} - 1| > 1/N.$$

For $x > 0$, the mean-value theorem gives

$$\ell n\ x = \xi^{-1}(x - 1),$$

where ξ lies between x and 1. Hence

$$\ell n|\alpha|^2 \geq N^{-1}\min(1, |\alpha|^{-2}) \geq N^{-1}[1 + d^2(\alpha)H^2(\alpha)]^{-1},$$

and $\ell n|\alpha|$ is effectively bounded below. Next, for $M \in \underset{\sim}{Z}^+$, find an $r \in \underset{\sim}{Q}$, $r > 0$, such that

$$|\alpha\bar{\alpha} - r| < 1/2MN.$$

Then by (11.1.14) again,

$$|\ell n\ \alpha\bar{\alpha} - \ell n\ r| \leq (2MN)^{-1}(\min(\alpha\bar{\alpha},r))^{-1}$$
$$< (2MN)^{-1}(2N) = 1/M$$

since $r \geq \alpha\bar{\alpha} - (2MN)^{-1}$. □

Lemma 11.1.10. If the positive real number s is effectively approximable, then it is effectively bounded below.

Proof. Since s is effectively approximable, we can obtain its decimal expansion to as many places as desired. If $s \geq 1$, there is no problem. Otherwise

$$s = .000\ldots 0a_1\ldots \qquad (a_1 \neq 0),$$

and .000...01 is a lower bound. □

The reader should note that being given a Cauchy sequence for a positive real number s does not imply that s is effectively bounded below.

A Diophantine inequality involving an algebraic number α is most satisfactory if all quantities involved are directly expressible in terms of $d(\alpha)$ and $H(\alpha)$. For this reason one should try to avoid whenever possible the use of Lemma 11.1.10 for producing lower bounds. However, the bounds obtained by this lemma are at any rate more satisfactory (from the constructive point of view) than those produced by the nonconstructive argument of Wirsing's theorem.

We now see why the stricter concept of explicitly bounded below was introduced in Definition 3.2.1. For example, in the conclusion of Lemma 11.1.9 we may replace "effectively bounded below" by "explicitly bounded below".

For later reference we make the following definition:

<u>Definition 11.1.4</u>. A vector $\underline{v} = (v_1,\ldots,v_n) \in \underset{\sim}{C}^n$ is said to be effectively approximable if v_i is effectively approximable for $1 \leq i \leq n$. It is said to be effectively bounded below if $\|\underline{v}\|^2 = |v_1|^2 + \cdots + |v_n|^2$ is effectively bounded below.

Our next goal is to show (Theorem 11.1.3) that if $\alpha,\beta \in \underset{\sim}{A}$ are effectively given, then we can decide whether or not $\beta \in \underset{\sim}{K} = \underset{\sim}{Q}(\alpha)$.

<u>Definition 11.1.5</u>. If $\beta \in \underset{\sim}{K} = \underset{\sim}{Q}(\alpha)$ and $d(\underset{\sim}{K}/\underset{\sim}{Q}) = n$, let $\sigma_1,\ldots,\sigma_n$ be the isomorphisms of $\underset{\sim}{K}$ into $\underset{\sim}{C}$ that fix $\underset{\sim}{Q}$. We define the norm and trace of β by

$$N_{\underset{\sim}{K}/\underset{\sim}{Q}}(\beta) = N(\beta) = \sigma_1(\beta)\cdots\sigma_n(\beta)$$

and

$$Tr_{\underset{\sim}{K}/\underset{\sim}{Q}}(\beta) = Tr(\beta) = \sigma_1(\beta) + \cdots + \sigma_n(\beta),$$

respectively.

Note that for $a,b \in \underset{\sim}{Q}$ and $\beta_1,\beta_2 \in \underset{\sim}{K}$ we have $N(a\beta_1\beta_2) = a^n N(\beta_1)N(\beta_2)$, $Tr(a\beta_1 + b\beta_2) = aTr(\beta_1) + bTr(\beta_2)$, and $N(\beta_1), Tr(\beta_1) \in \underset{\sim}{Q}$. Moreover, if $b_m z^m + \cdots + b_0$ is the minimal polynomial of β, then

$$N(\beta) = \{(-1)^m b_0/b_m\}^{n/m}$$

and

$$\mathrm{Tr}(\beta) = (n/m)(-b_{m-1}/b_m),$$

so the norm and trace of β are effectively given whenever α and β are effectively given. Furthermore, if β is an algebraic integer, then its norm and trace are rational integers.

Definition 11.1.6. Let $\alpha \in \underset{\sim}{A}$, $d(\alpha) = n$, and $\underset{\sim}{K} = \underset{\sim}{Q}(\alpha)$. We say that $\omega_1,\ldots,\omega_n$ are a basis for $\underset{\sim}{K}/\underset{\sim}{Q}$ if every $\omega_i \in \underset{\sim}{K}$ and if, for every $\beta \in \underset{\sim}{K}$,

$$\beta = r_1\omega_1 + \cdots + r_n\omega_n, \tag{11.1.14}$$

where $r_i \in \underset{\sim}{Q}$ for $1 \leq i \leq n$. We say that $\omega_1,\ldots,\omega_n$ are an integral basis for $\underset{\sim}{K}/\underset{\sim}{Q}$ if every ω_1 is an algebraic integer of $\underset{\sim}{K}$ and if, for every algebraic integer $\beta \in \underset{\sim}{K}$, we have

$$\beta = a_1\omega_1 + \cdots + a_n\omega_n, \tag{11.1.15}$$

where $a_i \in \underset{\sim}{Z}$ for $1 \leq i \leq n$.

It is clear that the r_i and a_i in (11.1.14) and (11.1.15) are unique.

Definition 11.1.7. Let $\omega_1,\ldots,\omega_n$ be a basis.

(i) We say that $\omega_1^*,\ldots,\omega_n^*$ are a basis dual to it (or a dual basis) if

$$\mathrm{Tr}(\omega_i\omega_j^*) = \delta_{ij} \qquad (1 \leq i,j \leq n),$$

where δ_{ij} is 1 if $i = j$ and 0 otherwise.

(ii) We say that $\Delta = \Delta_\omega = \Delta(\omega_1,\ldots,\omega_n) = [\det \omega_i^{(j)}]^2$ is its discriminant, where $\omega_i^{(1)},\ldots,\omega_i^{(n)}$ are the conjugates of ω_i,

$1 \leq i \leq n$. Note that

$$\Delta = \det[\mathrm{Tr}(\omega_i\omega_j)] = \begin{vmatrix} \omega_1^{(1)} & \cdots & \omega_1^{(n)} \\ \vdots & & \vdots \\ \omega_n^{(1)} & \cdots & \omega_n^{(n)} \end{vmatrix} \cdot \begin{vmatrix} \omega_1^{(1)} & \cdots & \omega_n^{(1)} \\ \vdots & & \vdots \\ \omega_1^{(n)} & \cdots & \omega_n^{(n)} \end{vmatrix}.$$

Lemma 11.1.11. Let $\omega_1,\ldots,\omega_n$ and $\theta_1,\ldots,\theta_n$ be bases for $\mathbb{K}/\mathbb{Q}$. Then

$$\omega_i = \sum_{j=1}^{n} b_{ij}\theta_j \qquad (1 \leq i \leq n), \tag{11.1.16}$$

where $b_{ij} \in \mathbb{Q}$, and

$$\Delta_\omega = (\det b_{ij})^2 \Delta_\theta \neq 0. \tag{11.1.17}$$

If the $\omega_1,\ldots,\omega_n$ are all algebraic integers, then Δ_ω is a nonzero integer.

Proof. Equation (11.1.17) is an immediate consequence of (11.1.16), which follows from Definition 11.1.6. Next, it is clear from (11.1.17) that if Δ_θ vanishes for any basis θ, then it vanishes for all bases ω. But Δ_ω does not vanish for the basis $1,\alpha,\ldots,\alpha^{n-1}$ because the determinant in this case is Vandermondian. That Δ_ω is an integer when $\omega_1,\ldots,\omega_n$ are algebraic integers follows from the symmetric function theorem: interchanging two conjugates of ω_i will change $\det \omega_i^{(j)}$ by -1, but Δ_ω is the square of this determinant. □

Lemma 11.1.12. A basis $\omega_1, \ldots, \omega_n$ of algebraic integers is an integral basis if and only if $|\Delta_\omega|$ is minimal.

Proof. Say $\omega_1, \ldots, \omega_n$ are an integral basis for K/Q and let $\theta_1, \ldots, \theta_n$ be any basis consisting of algebraic integers. Then

$$\theta_i = \sum_{j=1}^{n} a_{ij}\omega_j \qquad (1 \leq i \leq n),$$

where $a_{ij} \in Z$ and $\det a_{ij} \neq 0$ (by Lemma 11.1.11). Hence

$$\Delta_\theta = (\det a_{ij})^2 \Delta_\omega$$

and $|\Delta_\omega| \leq |\Delta_\theta|$. On the other hand, suppose $|\Delta_\omega|$ is minimal but that, for some algebraic integer $\beta \in K$,

$$\beta = a_1\omega_1 + \cdots + a_n\omega_n, \tag{11.1.18}$$

where $a_i \notin Z$ for some i, $1 \leq i \leq n$. On altering β by an appropriate integral multiple of ω_i and renumbering the ω_i, we may suppose that $i = 1$ and $0 < a_1 < 1$. By means of row and column operations it then follows that

$$\begin{aligned}|\Delta(\beta, \omega_2, \ldots, \omega_n)| &= |\Delta(a_1\omega_1, \omega_2, \ldots, \omega_n)| \\ &= a_1^2 |\Delta(\omega_1, \ldots, \omega_n)| < |\Delta_\omega|,\end{aligned}$$

a contradiction. □

It is immediate from Lemma 11.1.12 that integral bases for K/Q always exist. For $K \neq Q$ it can easily be shown that there are infinitely many and that any two of them are related by a unimodular matrix. We shall see now that an integral basis for K/Q

can be effectively found.

Lemma 11.1.13. If $\theta_1,\ldots,\theta_n$ are a basis and the θ_i are effectively given, then we can determine effectively a dual basis $\theta_1^*,\ldots,\theta_n^*$.

Proof. Define

$$\theta_i^* = \sum_{j=1}^{n} a_{ij}\theta_j \qquad (1 \le i \le n),$$

where the $a_{ij} \in \underset{\sim}{Q}$ will be determined later. We want

$$\mathrm{Tr}(\theta_i^*\theta_k) = \sum_{j=1}^{n} a_{ij}\ \mathrm{Tr}(\theta_j\theta_k) = \delta_{ik}.$$

This is a system of n equations in n unknowns with rational coefficients whose determinant does not vanish by Lemma 11.1.11. By Lemma 11.1.8 and the comment after Definition 11.1.5, the coefficients are effectively given. Hence we can effectively solve for the $a_{ij} \in \underset{\sim}{Q}$ (use Cramer's rule) and thus obtain the θ_i^*. □

Lemma 11.1.14. A dual basis is a basis.

Proof. Let $\theta_1^*,\ldots,\theta_n^*$ be a basis dual to $\theta_1,\ldots,\theta_n$. It suffices to show that $\theta_1^*,\ldots,\theta_n^*$ are linearly independent over $\underset{\sim}{Q}$. But if

$$a_1\theta_1^* + \cdots + a_n\theta_n^* = 0 \qquad (a_i \in \underset{\sim}{Q}),$$

then

$$a_i = a_1\ \mathrm{Tr}(\theta_i\theta_1^*) + \cdots + a_n\ \mathrm{Tr}(\theta_i\theta_n^*) = 0.$$

□

<u>Theorem 11.1.1</u>. If $\alpha \in \underset{\sim}{A}$ is effectively given, then an integral basis for $\underset{\sim}{K} = \underset{\sim}{Q}(\alpha)$ can be effectively found.

<u>Proof</u>. Let $d(\underset{\sim}{K}/\underset{\sim}{Q}) = n$. Without loss of generality we may suppose that α is an algebraic integer. Clearly $(\theta_1, \ldots, \theta_n) = (1, \alpha, \ldots, \alpha^{n-1})$ is a basis for $\underset{\sim}{K}$ consisting of algebraic integers. Find the dual basis $\theta_1^*, \ldots, \theta_n^*$. Let $\beta \in \underset{\sim}{K}$ be an algebraic integer. Then

$$\beta = \sum_{j=1}^{n} b_j \theta_j^* \qquad (b_j \in \underset{\sim}{Q}).$$

The last comment after Definition 11.1.5 shows that in fact $b_j \in \underset{\sim}{Z}$ for $1 \leq j \leq n$ since

$$b_j = \mathrm{Tr}(\theta_j \beta).$$

Hence if $\omega_1, \ldots, \omega_n$ are an integral basis,

$$\omega_i = \sum_{j=1}^{n} a_{ij} \theta_j^* \qquad (a_{ij} \in \underset{\sim}{Z}). \tag{11.1.19}$$

By means of row operations we can show that there exists an integral basis $\omega_1, \ldots, \omega_n$ [possibly different from the one of (11.1.19)] such that

$$\begin{aligned} \omega_1 &= a_{11}\theta_1^* + a_{12}\theta_2^* + \cdots + a_{1n}\theta_n^* \\ \omega_2 &= \qquad\quad\;\; a_{22}\theta_2^* + \cdots + a_{2n}\theta_n^* \\ &\;\;\vdots \qquad\qquad\qquad\qquad \ddots \qquad \vdots \\ \omega_n &= \qquad\qquad\qquad\qquad\qquad\;\; a_{nn}\theta_n^* \end{aligned} \tag{11.1.20}$$

where $a_{ij} \in \underset{\sim}{Z}$, and $a_{ij} = 0$ for $i > j$, and $0 \leq a_{ij} \leq |a_{jj}|$ for $i \leq j$. Now, by Lemma 11.1.12,

$$(11.1.21) \quad (a_{11} \cdots a_{nn})^2 |\Delta(\theta_1^*, \ldots, \theta_n^*)| = |\Delta(\omega_1, \ldots, \omega_n)| \leq |\Delta(\theta_1, \ldots, \theta_n)|.$$

Inequality (11.1.21) gives us an effective upper bound on $(a_{11} \cdots a_{nn})^2$. Hence there are only a finite number of possibilities for the matrix a_{ij} in (11.1.20). List all of them for which the corresponding ω_i are algebraic integers and for which $a_{11} \cdots a_{nn} \neq 0$. Of these, find one for which $(a_{11} \cdots a_{nn})^2$ is minimal. This must be an integral basis.

□

Definition 11.1.8. If $\omega_1, \ldots, \omega_n$ are an integral basis for $\underset{\sim}{K}/\underset{\sim}{Q}$, we say that $\Delta(\omega_1, \ldots, \omega_n)$ is the discriminant of $\underset{\sim}{K}$.

It is clear that the discriminant of $\underset{\sim}{K}$ can be effectively calculated.

Theorem 11.1.2. Let $\underset{\sim}{K} = \underset{\sim}{Q}(\alpha)$ and $d(\underset{\sim}{K}/\underset{\sim}{Q}) = n$, where α is effectively given. For $M \in \underset{\sim}{Z}^+$, let $S = S(M;\underset{\sim}{K})$ denote the set of all algebraic integers β such that $\beta \in \underset{\sim}{K}$ and

$$(11.1.22) \qquad |\beta_i| < M \qquad (1 \leq i \leq n),$$

where $\beta_1 = \beta, \beta_2, \ldots, \beta_n$ are the conjugates of β. Then we can effectively find a finite set T of algebraic integers such that

$$S \subseteq T.$$

Proof. Let $\omega_1,\ldots,\omega_n$ be an integral basis and $\omega_1^*,\ldots,\omega_n^*$ its dual. We can effectively find some $B \in \underset{\sim}{Z}^+$ that is an upper bound for the absolute values of all conjugates of elements of the dual basis. For a number β satisfying (11.1.22), write

$$\beta = b_1\omega_1 + \cdots + b_n\omega_n,$$

where $b_i \in \underset{\sim}{Z}$, $1 \leq i \leq n$. Then

$$|b_i| = |Tr(\omega_i^*\beta)| \leq nBM,$$

and the result follows. □

Theorem 11.1.3. If $\alpha,\beta \in \underset{\sim}{A}$ are effectively given, then we can decide whether or not $\beta \in \underset{\sim}{K} = \underset{\sim}{Q}(\alpha)$.

Proof. It is enough to consider the case in which β is an algebraic integer. Find an $M \in \underset{\sim}{Z}^+$ such that $M > d(\beta)H(\beta)$ and construct the set $T \supseteq S(M;\underset{\sim}{K})$ of Theorem 11.1.2. Then $\beta \in \underset{\sim}{K}$ if and only if $\beta \in T$. □

Lemma 11.1.15. Let $\alpha_1,\ldots,\alpha_n \in \underset{\sim}{A}$. If we are given an $F(x_1,\ldots,x_n) \in \underset{\sim}{Z}[x_1,\ldots,x_n]$ and upper bounds for the degrees and lengths of the α_i, then we can effectively find an $N \in \underset{\sim}{Z}^+$ such that either

$$F(\alpha_1,\ldots,\alpha_n) = 0 \tag{11.1.23}$$

or

$$|F(\alpha_1,\ldots,\alpha_n)| > 1/N. \tag{11.1.24}$$

Proof. Since $F(\alpha_1,\ldots,\alpha_n)$ is a sum of products of algebraic numbers, one simply applies Lemma 4.1.5(ii) and (iii) repeatedly; the result then follows from Lemma 3.2.1. Alternatively, use Lemma 4.1.7.

□

Lemma 11.1.16. Let $F(x_1,\ldots,x_n) \in \underset{\sim}{Z}[x_1,\ldots,x_n]$ and assume that $\underline{u} = (u_1,\ldots,u_n)$ and $\underline{v} = (v_1,\ldots,v_n)$ are complex vectors such that $\underline{u}$ is effectively bounded above and $\|\underline{u} - \underline{v}\| \leq 1$. Then we can effectively find an $M \in \underset{\sim}{Z}^+$ such that

$$|F(\underline{u}) - F(\underline{v})| \leq M\|\underline{u} - \underline{v}\|.$$

(Here M depends only on F and the upper bound for $\underline{u}$.)

Proof. Let the Taylor series of $F = F(x_1,\ldots,x_n)$ about $\underline{v}$ be

$$F = \sum C_{i_1\cdots i_n}(x_1 - v_1)^{i_1}\cdots(x_n - v_n)^{i_n}.$$

Then $F(\underline{u}) - F(\underline{v})$ has the above form, with x_i replaced by u_i and $C_{0,\ldots,0}$ replaced by 0. Since the (finitely many) $C_{i_1\cdots i_n}$ can be expressed as derivatives of F evaluated at $\underline{v}$, and $|v_i| \leq |u_i| + 1$, we can easily find an $M \in \underset{\sim}{Z}^+$ that bounds the sum of their absolute values. Hence

$$|F(\underline{u}) - F(\underline{v})| \leq M \max|u_i - v_i| \leq M\|u - v\|.$$

□

Under the assumptions of Lemma 11.1.15 we shall derive several corollaries, the first of which is a generalization of Liouville's theorem.

Corollary 11.1.1. Let $\underline{u} = (u_1,\ldots,u_n)$ and $\mathcal{A} = (\alpha_1,\ldots,\alpha_n)$. If $F(\underline{u}) = 0$ but $F(\mathcal{A}) \neq 0$, then we can effectively find an

$N_1 \in \underset{\sim}{Z}^+$ such that

$$\|\underline{u} - \mathcal{A}\| \geq 1/N_1.$$

Proof. We can assume that $\|\underline{u} - \mathcal{A}\| \leq 1$, and hence $\underline{u}$ is effectively bounded above. Thus, by Lemmas 11.1.15 and 11.1.16,

$$1/N < |F(\mathcal{A})| = |F(\underline{u}) - F(\mathcal{A})| < M\|\underline{u} - \mathcal{A}\|.$$

□

Corollary 11.1.2. Let $\beta_1,\ldots,\beta_n$ be algebraic numbers that are effectively given and set $\mathcal{B} = (\beta_1,\ldots,\beta_n)$. Then we can effectively determine whether or not $F(\mathcal{B}) = 0$.

Proof. By Lemma 11.1.15, we can effectively find an N so that either $F(\mathcal{B}) = 0$ or $|F(\mathcal{B})| > 1/N$. By Lemma 11.1.5, we can effectively find $r_1,\ldots,r_n \in \underset{\sim}{Q}(i)$ such that

$$\|\underline{r} - \mathcal{B}\| \leq 1/3NM,$$

where $\underline{r} = (r_1,\ldots,r_n)$ and M is the integer (depending only on F and $\mathcal{B}$) given by Lemma 11.1.16. Now

$$|F(\underline{r}) - F(\mathcal{B})| \leq M\|\underline{r} - \mathcal{B}\| \leq 1/3N.$$

Thus if $F(\mathcal{B}) = 0$, we have $|F(\underline{r})| \leq 1/3N$, whereas if $F(\mathcal{B}) \neq 0$, we have

$$|F(\underline{r})| \geq |F(\mathcal{B})| - |F(\mathcal{B}) - F(\underline{r})| > 1/N - 1/3N > 1/3N.$$

□

Corollary 11.1.3. Given a polynomial $P(z) \in \underset{\sim}{Z}[z]$ and a planar curve

$$\mathcal{C} = \{(x,y)\,|\,F(x,y) = 0\},$$

where $F(x,y) \in \underset{\sim}{Z}[x,y]$, we can effectively determine all the roots α of $P(z)$ that lie on $\mathcal{C}$.

Proof. Let $\beta_1 = \mathrm{Re}(\alpha)$, $\beta_2 = \mathrm{Im}(\alpha)$, and apply Corollary 11.1.2. □

In particular, if we take $F(x,y) = y$, we have an algorithm for computing the number of real roots of a polynomial with integer coefficients. This is usually done by Sturm's theorem (see van der Waerden [123], pp. 218-222), which we have avoided here. However, Sturm's theorem is fundamental in much of constructive mathematics and forms the basis for the very general algorithm of Tarski [121].

Exercises

11.1.1. Show that if s_1 and s_2 are effectively approximable, then so are $-s_1$, $s_1 + s_2$, and $s_1 s_2$. (Hint: For $s_1 + s_2$ find r_1 and r_2 in $\underset{\sim}{Q}(i)$ such that $|s_1 - r_1|$ and $|s_2 - r_2|$ are both less than $1/2N$.)

11.1.2. Prove Lemmas 11.1.4 and 11.1.5 under a weaker assumption than $P(z) \in \underset{\sim}{Z}[z]$. (Hint: Try $P(z)$ monic, with all coefficients effectively approximable.)

11.1.3. Let $m \in \underset{\sim}{Z}^+$. Show that if $\alpha \in \underset{\sim}{A}$ is effectively given, then so are all m roots of the equation $z^m = \alpha$.

11.1.4. Define "effectively bounded above" and show that it is implied by "effectively approximable".

11.1.5. If $\alpha_1, \alpha_2, \beta \in \underset{\sim}{A}$ and α_1, α_2 have the same minimal polynomial, do $\alpha_1\beta$, $\alpha_2\beta$ have the same minimal polynomial?

11.1.6. Find integral bases for the real fields $\underset{\sim}{Q}(12^{1/3})$ and $\underset{\sim}{Q}(28^{1/3})$.

11.1.7. Given a basis $\omega_1,\ldots,\omega_n$, is its dual basis unique? What can you say about the dual basis of the dual basis?

11.1.8. Show that the $\Delta(\theta_1^*,\ldots,\theta_n^*)$ of (11.1.21) is effectively bounded below. Is it explicitly bounded below?

*11.1.9. If the discriminant of $\underset{\sim}{K}$ is 1, does $\underset{\sim}{K} = \underset{\sim}{Q}$?

11.1.10. If $\alpha,\beta \in \underset{\sim}{A}$ are effectively given, show that one can effectively decide whether or not $\mathrm{Re}(\alpha) < \mathrm{Re}(\beta)$.

11.1.11 (See Corollary 11.1.3.). If $\underset{\sim}{C} - \mathcal{C} = \bigcup_j R_j$, where each R_j is an open connected set, can one effectively determine $r_j \in \underset{\sim}{Q}(i)$ such that $r_j \in R_j$? Can one effectively determine how many roots of $P(z)$ lie in each R_i?

11.1.12. Say $\alpha \in \underset{\sim}{A}$ is effectively given. Show that one can effectively find an $m \in \underset{\sim}{Z}$ such that, for every algebraic integer $\beta \in \underset{\sim}{Q}(\alpha)$,

$$m\beta = a_{n-1}\alpha^{n-1} + \cdots + a_1\alpha + a_0$$

with $a_i \in \underset{\sim}{Z}$, $0 \le i \le n - 1$.

11.1.13. Given an $\alpha \in \underset{\sim}{A}$, consider all possible representations of α in the form

$$\alpha = \frac{a_1\omega_1 + \cdots + a_n\omega_n}{b_1\omega_1 + \cdots + b_n\omega_n},$$

where $\omega_1,\ldots,\omega_n$ are some fixed integral basis for $\underset{\sim}{Q}(\alpha)$, the $a_i,b_i \in \underset{\sim}{Z}$ for $1 \le i \le n$, and $(a_1,\ldots,a_n,b_1,\ldots,b_n) = 1$. Let

$h = \max(|a_1|,\ldots,|a_n|,|b_1|,\ldots,|b_n|)$ and then define $h^* = h^*(\alpha;\omega_1,\ldots,\omega_n)$ to be the smallest value that h takes in any such representation. Does h^* depend effectively on α and $\omega_1,\ldots,\omega_n$? (In Gelfond [39], p. 7, the quantity h^* is called the <u>measure</u> of α.)

<u>11.1.14</u>. If the algebraic numbers $\alpha_1,\ldots,\alpha_n$ are effectively given, can we effectively determine whether or not they are linearly independent over $\underset{\sim}{Q}$?

11.2. Some Geometry of Numbers

Many problems of Diophantine approximation have the following form: "given a real-valued function $f(x_1,\ldots,x_n)$ and a number $b > 0$, is there a vector $(a_1,\ldots,a_n)$ with integral components such that $|f(a_1,\ldots,a_n)| < b$?" This can be phrased geometrically: "When will a point set in Euclidean space [given by $|f(x_1,\ldots,x_n)| < b$] contain a lattice point?" Theorems 11.2.1 and 11.2.2 are basic in the study of this question.

<u>Definition 11.2.1</u>. An <u>n-dimensional lattice</u> Λ is a subset of $\underset{\sim}{R}^n$ of the form

$$\{a_1\underline{v}_1 + \cdots + a_n\underline{v}_n \mid a_i \in \underset{\sim}{Z} \text{ and } \underline{v}_1,\ldots,\underline{v}_n \text{ are n fixed linearly independent vectors of } \underset{\sim}{R}^n\}. \tag{11.2.1}$$

If $\underline{w}_1,\ldots,\underline{w}_n \in \Lambda$ have the property that, for all $\underline{w} \in \Lambda$,

$$\underline{w} = a_1\underline{w}_1 + \cdots + a_n\underline{w}_n,$$

where $a_i \in \underset{\sim}{Z}$, $1 \le i \le n$, then we say that $\underline{w}_1,\ldots,\underline{w}_n$ are a <u>basis</u> of Λ.

For example, $\underline{v}_1,\ldots,\underline{v}_n$ are a basis of the lattice given by (11.2.1). The concept of a lattice basis is very similar to that of an integral basis (see Definition 11.1.6).

Definition 11.2.2. Consider $\underset{\sim}{R}^n$ as consisting of column vectors. If $A = (a_{ij})$ is a real nonsingular $n \times n$ matrix and Λ is given by (11.2.1), we set

$$A\Lambda = \{A\underline{w} \mid \underline{w} \in \Lambda\}.$$

It is easy to verify that $A\Lambda$ is also a lattice and that if $\underline{w}_1,\ldots,\underline{w}_n$ are a basis for Λ, then $A\underline{w}_1,\ldots,A\underline{w}_n$ are a basis for $A\Lambda$. In the following we denote by Λ_0 the lattice of all points in $\underset{\sim}{R}^n$ with integral coordinates. The vectors $\underline{e}_1 = (1,0,\ldots,0)$, $\underline{e}_2 = (0,1,0,\ldots,0),\ldots,\underline{e}_n = (0,\ldots,0,1)$ form a basis for Λ_0. If $\underline{v}_1,\ldots,\underline{v}_n$ is a basis for Λ and A is the real nonsingular $n \times n$ matrix such that $\underline{v}_i = A\underline{e}_i$ for $1 \le i \le n$, then $\Lambda = A\Lambda_0$.

Lemma 11.2.1. If $\underline{v}_1,\ldots,\underline{v}_n$ and $\underline{w}_1,\ldots,\underline{w}_n$ are bases of Λ, then

$$\underline{v}_i = \sum_{j=1}^{n} a_{ij}\underline{w}_j \qquad (1 \le i \le n), \tag{11.2.2}$$

where $A = (a_{ij})$ is an integral matrix with $\det A = \pm 1$.

Proof. Let I be the $n \times n$ identity matrix. Write

$$\underline{w}_j = \sum_{k=1}^{n} b_{ij}\underline{v}_k \qquad (1 \le j \le n)$$

and set $B = (b_{jk})$. From (11.2.2) we find that $I = AB$, so $1 = (\det A)(\det B)$. Since A and B have integral entries, the result follows.

□

From Lemma 11.2.1 we see that $|\det(\underline{v}_1,\ldots,\underline{v}_n)|$ has the same value for any basis of Λ.

Definition 11.2.3. We define the determinant of the lattice Λ as

$$d(\Lambda) = \det(\Lambda) = |\det(\underline{v}_1,\ldots,\underline{v}_n)|,$$

where $\underline{v}_1,\ldots,\underline{v}_n$ are any basis of Λ.

It is clear that $d(A\Lambda) = |\det A|d(\Lambda)$ for any lattice Λ. In the next lemma and its corollary we follow the treatment of Lekkerkerker [57], pp. 18 and 19. A set of points (i.e., vectors) in $\underset{\sim}{R}^n$ is said to be discrete if it has no limit points. As in Definition 11.1.4, we denote the distance of the vector $\underline{v} \in \underset{\sim}{R}^n$ from the origin by $\|\underline{v}\|$.

Lemma 11.2.2. The set Λ is an n-dimensional lattice if and only if it is a discrete subgroup of $\underset{\sim}{R}^n$ that contains n linearly independent elements.

Proof. If Λ is a lattice, write $\Lambda = A\Lambda_0$, where A is a non-singular $n \times n$ matrix. Since Λ_0 is discrete and $\Lambda_0 \doteq A^{-1}\Lambda$, it is clear that Λ is discrete. On the other hand, say Λ is a discrete subgroup of $\underset{\sim}{R}^n$ with n linearly independent elements $\underline{w}_1,\ldots,\underline{w}_n$. Choose $\underline{v}_1 \in \Lambda$ so that (i) $\underline{v}_1 \neq 0$; (ii) $\underline{v}_1 = t_1\underline{w}_1$, where $0 \leq t_1 \leq 1$; and (iii) $\|\underline{v}_1\|$ is minimal. In general, if $\underline{v}_1,\ldots,\underline{v}_{i-1}$ have been chosen already $(1 \leq i < n + 1)$, we pick a vector $\underline{v}_i \in \Lambda$ (perhaps $\underline{w}_i$) such that (i) $\underline{v}_i$ is not in the space $L(i - 1) \subseteq \underset{\sim}{R}^n$ spanned by $\underline{v}_1,\ldots,\underline{v}_{i-1}$; (ii) $\underline{v}_i$ lies in the i-dimensional parallelopiped

$$(11.2.3)\qquad P_i = \{t_1\underline{v}_1 + \cdots + t_{i-1}\underline{v}_{i-1} + t_i\underline{w}_i \mid 0 \leq t_j \leq 1,\ 1 \leq j \leq i\};$$

and (iii) $\|\underline{v}_i - L(i-1)\| = \inf_{\underline{v}\in L(i-1)}\|\underline{v}_i - \underline{v}\|$ is minimal. [Since Λ is discrete, a vector $\underline{v}_i$ minimizing the expression of (iii) exists.] Clearly the $\underline{v}_i$ are linearly independent. For $\underline{v} \in \Lambda$, let

$$(11.2.4)\qquad \underline{v} = a_1\underline{v}_1 + \cdots + a_n\underline{v}_n,$$

where $a_i \in \underset{\sim}{R}$. Since Λ is a group,

$$\underline{v}' = \{a_1\}\underline{v}_1 + \cdots + \{a_n\}\underline{v}_n \in \Lambda,$$

where $\{a\} = a - [a]$ and $[a]$ denotes the greatest integer in a. Clearly $\underline{v}' \in P_n$. If $\{a_n\} \neq 0$, then

$$\begin{aligned}\|\underline{v}' - L(n-1)\| &= \|\{a_n\}\underline{v}_n - \{a_n\}L(n-1)\| \\ &= |\{a_n\}|\|\underline{v}_n - L(n-1)\| \\ &< \|\underline{v}_n - L(n-1)\|.\end{aligned}$$

This, however, contradicts the choice of $\underline{v}_n$. Now $a_{n-1} \neq [a_{n-1}]$ leads to the same type of contradiction, and so forth. Hence the a_i are integers, and so by (11.2.4) Λ is a lattice with basis $\underline{v}_1,\ldots,\underline{v}_n$. □

<u>Corollary 11.2.1</u>. If $\underline{w}_1,\ldots,\underline{w}_n$ are linearly independent points of Λ, we can choose a basis $\underline{v}_1,\ldots,\underline{v}_n$ of Λ such that

(11.2.5) $$\underline{w}_j = a_{1j}\underline{v}_1 + \cdots + a_{jj}\underline{v}_j \qquad (1 \leq j \leq n),$$

where $a_{ij} \in \underset{\sim}{Z}$.

Proof. Find the basis $\underline{v}_1,\ldots,\underline{v}_n$ exactly as in the proof of Lemma 11.2.2. Then it is clear that $\underline{v}_j$ is in the space spanned by $\underline{v}_1,\ldots,\underline{v}_{j-1}$ and $\underline{w}_j$. Since $\underline{v}_1,\ldots,\underline{v}_{j-1}$, $\underline{v}_j$ are linearly independent, $\underline{w}_j$ must be in the space spanned by $\underline{v}_1,\ldots,\underline{v}_j$. Hence (11.2.5) holds with the $a_{ij} \in \underset{\sim}{R}$. Since $\underline{w}_j \in \Lambda$,

$$\underline{w}_j = b_{1j}\underline{v}_1 + \cdots + b_{nj}\underline{v}_n,$$

where $b_{ij} \in \underset{\sim}{Z}$. Since the $\underline{v}_i$ are linearly independent, we see that $a_{ij} = b_{ij} \in \underset{\sim}{Z}$ for $1 \leq i \leq j$.

□

If S is a (Lebesgue) measurable set in $\underset{\sim}{R}^n$, we let V(S) denote its measure. The symbol ϕ shall denote the empty set.

Theorem 11.2.1 (Blichfeldt). If $V(S) > 1$, then there are points $\underline{x}$ and $\underline{y}$ in S such that $\underline{x} - \underline{y} \in \Lambda_0$ and $\underline{x} \neq \underline{y}$.

Proof. The intersection of S with a sufficiently large sphere centered at the origin will have measure greater than unity. It suffices to prove the theorem with S replaced by this bounded set. Let $\mathscr{A} = \{(x_1,\ldots,x_n) \mid 0 \leq x_i < 1\}$ and let $\underline{w}_1,\ldots,\underline{w}_N$ be the finite set of all lattice points $\underline{w} \in \Lambda_0$ such that

$$S(\underline{w}) = S \cap \{\mathscr{A} + \underline{w}\} \neq \phi.$$

Now $S'(\underline{w}_i) \equiv S(\underline{w}_i) - \underline{w}_i \subseteq \mathscr{A}$ for $1 \leq i \leq N$, and

$$\sum_{i=1}^{N} V[S'(\underline{w}_i)] = \sum_{i=1}^{N} V[S(\underline{w}_i)] = V(S) > 1.$$

Hence some two of the $S'(\underline{w}_i)$ overlap, so $\underline{x} - \underline{w}_i = \underline{y} - \underline{w}_j$ for $\underline{x},\underline{y} \in S$ and $i \neq j$.

□

It is easy to see that the conclusion of Blichfeldt's theorem is true if $V(S) \geq 1$ and S is compact (apply the above to λS, where $\lambda > 1$, then let $\lambda \to 1$, and use the fact that Λ_0 is discrete).

Definition 11.2.4. A set $S \subseteq \underset{\sim}{R}^n$ is said to be convex if $\underline{x},\underline{y} \in S$ implies that

$$\lambda\underline{x} + (1 - \lambda)\underline{y} \in S$$

for $0 \leq \lambda \leq 1$; that is, the line segment joining $\underline{x}$ and $\underline{y}$ is also in S. The set S is said to be centrally symmetric if $\underline{x} \in S$ implies that $-\underline{x} \in S$.

Theorem 11.2.2 (Minkowski). If $S \subseteq \underset{\sim}{R}^n$ is a convex centrally symmetric set with $V(S) > 2^n$, then S contains a point of Λ_0 other than 0.

Proof. We have $V(\frac{1}{2} S) > 1$, so there are $\underline{x},\underline{y} \in \frac{1}{2} S$ with $\underline{x} \neq \underline{y}$ and

$$\underline{x} - \underline{y} = \frac{1}{2} (2\underline{x} - 2\underline{y}) \in \Lambda_0$$

by Blichfeldt's theorem (Theorem 11.2.1). Now $2\underline{x} \in S$ and $2\underline{y} \in S$, so $-2\underline{y} \in S$ by the central symmetry and $[2\underline{x} + (-2\underline{y})]/2 \in S$ by the convexity.

□

It is easy to see that if S is compact, $V(S) > 2^n$ may be replaced by $V(S) \geq 2^n$.

Corollary 11.2.2. Let Λ be an n-dimensional lattice. If $S \subseteq \mathbb{R}^n$ is a convex centrally symmetric set with $V(S) > 2^n d(\Lambda)$, then S contains a point of Λ other than 0.

Proof. For some matrix A, we have $\Lambda = A\Lambda_0$. Now

$$V(A^{-1}S) = |\det A|^{-1}V(S) = d(\Lambda)^{-1}V(S) > 2^n,$$

so there is an $\underline{x} \in A^{-1}S \cap \Lambda_0$ with $\underline{x} \neq 0$. Clearly $A\underline{x} \in S \cap \Lambda$. □

As before, $V(S) > 2^n d(\Lambda)$ can be replaced by $V(S) \geq 2^n d(\Lambda)$ if S is compact.

Definition 11.2.5. For $\underline{v} \in \Lambda$, the ray of $\underline{v}$ is

$$\rho(\underline{v}) = \{t\underline{v} | t \text{ is real}\}.$$

The point $\underline{v} \in \Lambda$ is called primitive (or visible) if $\underline{v} = n\underline{w}$ for $n \in \mathbb{Z}$ and $\underline{w} \in \Lambda$ implies that $n = \pm 1$.

Lemma 11.2.3. If $\underline{w} \in \Lambda \cap \rho(\underline{v})$ and $\underline{v}$ is primitive, then $\underline{w} = m\underline{v}$ for some $m \in \mathbb{Z}$.

Proof. Since Λ is discrete, there is some $\underline{v}_0 \in \rho(\underline{v})$ with minimal positive distance from 0. Choose $n \in \mathbb{Z}$ so that $\|\underline{v} - n\underline{v}_0\| < \|\underline{v}_0\|$. Then $\underline{v} = n\underline{v}_0$. Hence $\underline{v} = \pm \underline{v}_0$ since $\underline{v}$ is primitive. Choose $m \in \mathbb{Z}$ so that $\|\underline{w} - m\underline{v}\| < \|\underline{v}\|$. Then $\underline{w} = m\underline{v}$. □

We now develop lattice theory in an effective manner.

Definition 11.2.6. The lattice Λ is effectively given if for any $M \in \mathbb{Z}^+$ we can effectively find a finite, effectively approximable, and effectively bounded below set $S \subseteq \Lambda$ such that if

(i) $\underline{v} \in \Lambda$

and

(ii) $\|\underline{v}\| < M$,

then $\underline{v} \in S$.

Lemma 11.2.4. Say $\underline{v} \in \Lambda$, $\underline{v} \neq 0$, and let Λ' denote the projection of Λ on a hyperplane H through the origin that is perpendicular to $\underline{v}$. Then Λ' is an (n - 1)-dimensional lattice. Moreover, if (i) $\underline{v}$ is primitive; (ii) $\|\underline{v}\| < N \in \underset{\sim}{Z}^+$, where N is given; and (iii) Λ is effectively given, then Λ' is effectively given.

Proof. The first part can be deduced easily from Lemma 11.2.2. We shall give a different proof. Without loss of generality, $\underline{v}$ is primitive. It follows from the proof of Lemma 11.2.2 that $\underline{v} = \underline{v}_1$ is part of a basis $\underline{v}_1,\ldots,\underline{v}_n$. For $\underline{w} \in \Lambda$, let $P(\underline{w})$ denote the projection of $\underline{w}$ on H. Clearly P is linear, so

(11.2.6) $$\Lambda' = \{P(a_1\underline{v}_1 + \cdots + a_n\underline{v}_n) \,|\, a_i \in \underset{\sim}{Z},\ 1 \leq i \leq n\}$$

$$= \{a_2P(\underline{v}_2) + \cdots + a_nP(\underline{v}_n) \,|\, a_i \in \underset{\sim}{Z},\ 2 \leq i \leq n\}.$$

Clearly $P(\underline{v}_2),\ldots,P(\underline{v}_n)$ are linearly independent since they generate a linear space of dimension (n - 1). Thus (11.2.6) shows that Λ' is an (n - 1)-dimensional lattice.

For the second part of the theorem let $M \in \underset{\sim}{Z}^+$. We want to find a finite, effectively approximable, and effectively bounded below set $S' \subseteq \Lambda'$ such that

(i) $\underline{w}' \in \Lambda'$

and

(ii) $\|\underline{w}'\| < M$

imply that $\underline{w}' \in S'$. For any such $\underline{w}'$, we have $\underline{w}' = \underline{w} + t\underline{v}$, where $\underline{w} \in \Lambda$ and t is real. We may suppose that $|t| \leq 1/2$. Hence $\|\underline{w}\| \leq M + N/2$. Therefore, since Λ is effective, we can find a finite, effectively approximable, and effectively bounded below set $S \subseteq \Lambda$ such that the elements of S' are projections of elements of S. Now we must show that the projection can be done effectively. If $\underline{w} \in \Lambda$ and

$$\xi = (\underline{w}\cdot\underline{v})/\|\underline{v}\|^2,$$

then $(\underline{w} - \xi\underline{v})\cdot\underline{v} = 0$, so $\underline{w}' = \underline{w} - \xi\underline{v}$ is the projection of $\underline{w}$. Since Λ is effective, $\underline{v}$ and $\underline{w}$ are effectively approximable and effectively bounded below, so by Lemma 11.1.1, the quantity ξ and hence $\underline{w}'$ are effectively approximable. If $\underline{w}' = 0$, then $\xi \in \underset{\sim}{Z}$, by Lemma 11.2.3. Since ξ is effectively approximable, we can calculate it exactly. Since Λ is effective, we can decide if it is actually the case that $\underline{w}' = \underline{w} - \xi\underline{v} = 0$ (recall the definition of "effectively bounded below"). If $\underline{w}' \neq 0$, it is effectively bounded below by Lemma 11.1.10.

□

It is important to have a proof of Lemma 11.2.4 that does not use Lemma 11.1.10. Let

$$\mu(\Lambda) = \min\{\|\underline{x}\| \mid \underline{x} \in \Lambda \text{ and } \underline{x} \neq 0\}.$$

<u>Proposition 11.2.1</u>. Let $\underline{v} \in \Lambda$, an n-dimensional lattice, and let Λ' be the projection of Λ on the hyperplane H perpendicular to $\underline{v}$. Then there is some $c_1 > 3/4$ such that

$$\|\underline{v}\|\mu(\Lambda') \geq c_1\mu(\Lambda)^2,$$

and this result is best possible in the sense that if $\underline{v} = (0,0,1)$, then there is a positive constant $c_2 > 0$ and infinitely many three-dimensional lattices Λ such that

$$\mu(\Lambda') \leq c_2\mu(\Lambda)^2.$$

<u>Proof</u>. The set Λ' is a lattice by Lemma 11.2.4. Let $\mu(\Lambda') = \|\underline{w}'\|$, where $\underline{w}'$ is the projection of $\underline{w} \in \Lambda$. Clearly $\underline{v}$ and $\underline{w}$ are linearly independent. Let L and A be the linear space and parallelogram generated by $\underline{v}$ and $\underline{w}$; we shall also write A for the area of the parallelogram. Now $\Lambda'' = L \cap \Lambda$ is easily seen to be a two-dimensional lattice by Lemma 11.2.2, and $d(\Lambda'') \leq A$. Hence, by Minkowski's theorem (Theorem 11.2.2),

$$\pi\mu(\Lambda)^2 \leq 2^2 d'(\Lambda'') \leq 4A.$$

Since $A = \|\underline{w}'\|\|\underline{v}\|$, the first inequality follows with $c_1 = \pi/4$. On the other hand, let Λ have the fundamental basis $(\varepsilon,0,\tau)$, $(0,1,0)$, $(0,0,1)$, where $0 < \varepsilon < 1$ and τ is a quadratic irrationality. Set $\underline{v} = (0,0,1)$. Then by Liouville's theorem there is a positive constant $c_3 > 0$ such that

$$\begin{aligned}\mu(\Lambda) &= \min_{n_i \in \mathbb{Z}} \{(n_1\varepsilon)^2 + n_2^2 + (n_1\tau + n_3)^2\}^{1/2} \\ &\geq \min_{n \in \mathbb{Z}} \{(n\varepsilon)^2 + (c_2/n)^2\}^{1/2} \\ &\geq (2c_3\varepsilon)^{1/2}.\end{aligned}$$

Since $\mu(\Lambda') = \varepsilon$, we conclude that

$$\mu(\Lambda') \leq (1/2c_3)\mu(\Lambda)^2.$$

□

Thus we no longer need Lemma 11.1.10 to establish the effectively bounded below property in the proof of Lemma 11.2.4. If we take $\tau = (\sqrt{5} - 1)/2$, it is not hard to show that $c_2 = (2 + \tau)^{-1}$, but this still leaves a considerable gap between c_1 and c_2. If $\Lambda = \Lambda_0$, we can take $c_1 = 1$ since (in the notation of the proof)

$$A^2 = \|\underline{v}\|^2\|\underline{w}\|^2 - (\underline{v}\cdot\underline{w})^2 \in \mathbb{Z}^+,$$

so $A \geq 1$.

<u>Lemma 11.2.5</u>. In the notation of Lemma 11.2.4, if $\underline{v} = \underline{v}_n$ is primitive and $\underline{v}_1, \ldots, \underline{v}_{n-1} \in \Lambda$ project into a basis $\underline{v}_1', \ldots, \underline{v}_{n-1}'$ of Λ', then $\underline{v}_1, \ldots, \underline{v}_{n-1}, \underline{v}_n$ are a basis of Λ. Also,

$$d(\Lambda) = \|\underline{v}_n\| d(\Lambda'). \tag{11.2.7}$$

<u>Proof</u>. We have $\underline{v}_i' = \underline{v}_i - t_i\underline{v}_n$, where $t_i \in \mathbb{R}$ for $1 \leq i \leq n - 1$. Hence, for every $\underline{w} \in \Lambda$, there are integers a_i and a real t such that

$$\underline{w} - t\underline{v}_n = \sum_{i=1}^{n} a_i(\underline{v}_i - t_i\underline{v}_n).$$

Thus

$$\underline{w} = a_1\underline{v}_1 + \cdots + a_{n-1}\underline{v}_{n-1} + u\underline{v}_n \qquad (u \in \underset{\sim}{R}).$$

By Lemma 11.2.3, we have $u \in \underset{\sim}{Z}$. Finally,

$$|\det(\underline{v}_1, \ldots, \underline{v}_{n-1}, \underline{v}_n)| = |\det(\underline{v}'_1, \ldots, \underline{v}'_{n-1}, \underline{v}_n)|$$
$$= \|\underline{v}_n\| \, |\det(\underline{v}'_1, \ldots, \underline{v}'_{n-1})|.$$

□

Corollary 11.2.3. Under the hypothesis of Lemma 11.2.4, if $d(\Lambda)$ is effectively bounded below, then so is $d(\Lambda')$.

Lemma 11.2.6. If Λ is effectively given, together with its dimension n, and also an $N \in \underset{\sim}{Z}^+$ such that $d(\Lambda) < N$, then we can effectively find an $M \in \underset{\sim}{Z}^+$ such that the sphere of radius M centered at the origin contains a basis of Λ.

Proof. We induct on $n = \dim \Lambda$. If $n = 1$, we obviously can take $M = N + 1$. Assume that this is true for effective lattices Γ with $\dim \Gamma = n - 1$. For an n-dimensional Λ find a finite, effectively approximable, and effectively bounded below set S containing all lattice points within a sphere of radius $\sqrt{n}\,(N + 1)^{1/n}$. This sphere contains a box of side $2(N + 1)^{1/n}$, and hence its volume exceeds $2^n N > 2^n d(\Lambda)$. So by Minkowski's theorem (Theorem 11.2.2) we find that S is nonempty. By Lemma 11.1.2 with $m = 2$ we can find a primitive $\underline{v}_n \in S$. Now consider the Λ' lattice of Lemma 11.2.4 with $\underline{v} = \underline{v}_n$. Here $d(\Lambda') =$

$d(\Lambda)/\|\underline{v}_n\| \leq N/\|\underline{v}_n\|$ by Lemma 11.2.5, and $\underline{v}_n$ is effectively bounded below, so we can find an $N' \in \underset{\sim}{Z}^+$ such that $d(\Lambda') < N'$. Hence by the induction hypothesis we can find an M' for Λ'. Consider the set S' of all points of Λ' within a sphere of radius M'. Let $\underline{v}_1', \ldots, \underline{v}_{n-1}'$ be a basis for Λ' from S'. Clearly

$$\underline{v}_i' = \underline{v}_i + t_i\underline{v}_n \qquad (1 \leq i \leq n-1) \tag{11.2.8}$$

for some $\underline{v}_i \in \Lambda$, where $|t_i| \leq 1/2$. From (11.2.8) and the bounds on $\underline{v}_i'$ and $\underline{v}_n$ we can easily find some $C \in \underset{\sim}{Z}^+$ such that $\|\underline{v}_i\| < C$ for $1 \leq i \leq n-1$. Since $\underline{v}_1, \ldots, \underline{v}_{n-1} \in \Lambda$ project into a basis of Λ', it follows from Lemma 11.2.5 that $\underline{v}_1, \ldots, \underline{v}_{n-1}, \underline{v}_n$ form a basis of Λ. Hence we can take any $M \in \underset{\sim}{Z}^+$ such that

$$M > \max(C, \|\underline{v}_n\|).$$

□

Lemma 11.2.7. If the n-dimensional lattice Λ is effectively given and also an $M \in \underset{\sim}{Z}^+$ such that the sphere of radius M centered at the origin contains a basis of Λ, then we can effectively find a basis for Λ.

Proof. Find a finite set S corresponding to M as in Definition 11.2.6. List all of its n element subsets $S_1, \ldots, S_u$. Find an $N \in \underset{\sim}{Z}^+$ such that $\|s\| \geq 1/N$ for all $s \in S$. Then the box

$$|x_i| \leq (2\sqrt{n}\,N)^{-1} \qquad (1 \leq i \leq n)$$

satisfies the hypotheses of Corollary 11.2.2, yet contains no non-zero lattice points. Hence

$$(\sqrt{n}\,N)^{-n} \leq 2^n d(\Lambda),$$

and we can find an $N_1 \in \underset{\sim}{Z}^+$ such that $d(\Lambda) > 1/N_1$. Now the numbers $|\det S_i|$, where $1 \leq i \leq u$, are effectively approximable; calculate them to within $1/3N_1$. These numbers are clearly a subset of

$$\{md(\Lambda)\,|m \geq 0,\ m \in \underset{\sim}{Z}\},$$

so any two of them are either equal or differ by at least $1/N_1$. Hence they can be ordered. Clearly those S_i with $|\det S_i|$ nonzero and minimal are the bases.

□

Corollary 11.2.4. If Λ is effectively given together with its dimension n and also an $N \in \underset{\sim}{Z}^+$ such that $d(\Lambda) < N$, then $d(\Lambda)$ is effectively bounded below, and we can effectively find a basis of Λ.

Proof. By Lemmas 11.2.6, 11.2.7, and the proof of Lemma 11.2.7.

□

Exercises

11.2.1. Show that in Corollary 11.2.1 the basis $\underline{v}_1,\ldots,\underline{v}_n$ may be chosen so that $a_{jj} > 0$ and $0 \leq a_{ij} < a_{jj}$ for $i < j$.

11.2.2. If Λ_1 and Λ are n-dimensional lattices and $\Lambda_1 \subseteq \Lambda$, we say that Λ_1 is a sublattice of Λ. Show that

$$I = [\Lambda : \Lambda_1] = d(\Lambda_1)/d(\Lambda),$$

where I is the index of the additive group Λ_1 in Λ.

11.2.3. Show that $\underline{v}_1,\ldots,\underline{v}_n$ is a basis of Λ if and only if

$$|\det(\underline{v}_1,\ldots,\underline{v}_n)| = d(\Lambda).$$

(Hint: Use Exercise 11.2.2.)

11.2.4. Deduce the first part of Lemma 11.2.4 directly from Lemma 11.2.2.

11.2.5. Show that the intersection of $\Lambda \subseteq \mathbb{R}^n$ with a subspace of $\mathbb{R}^n$ is an m-dimensional lattice for some $m \in \mathbb{Z}$, $0 \leq m \leq n$. (Hint: Use Lemma 11.2.2.)

11.2.6. If A is an $n \times n$ matrix of integers and Λ_0 is an n-dimensional lattice, show that $A\Lambda_0$ is an m-dimensional lattice for some $m \in \mathbb{Z}$ with $0 \leq m \leq n$. If A is given, can a basis for $A\Lambda_0$ be effectively found? (Hint: Use Lemma 11.2.2 for the first part.)

11.2.7. If Λ is an effective n-dimensional lattice and we are given an $M \in \mathbb{Z}^+$ such that the sphere of radius M centered at the origin contains n linearly independent points, show that we can effectively find a basis of Λ. (Hint: Examine the proof of Lemma 11.2.2.)

11.2.8. Say A is a real $n \times n$ matrix whose entries are effectively approximable and $N_1, N_2 \in \mathbb{Z}^+$ are so given that

$$N_1^{-1} < \|A\underline{x}\| < N_2$$

for all $\underline{x} \in \mathbb{R}^n$ such that $\|\underline{x}\| = 1$. Show that $A\Lambda_0$ is an effective lattice.

Now let K denote an open, convex, centrally symmetric set in $\mathbb{R}^n$. We define $\lambda_i = \lambda_i(K,\Lambda)$, the i-th successive minimum of K with respect to the lattice Λ, to be

$$\inf\{\lambda \mid \lambda > 0 \text{ and } \lambda K \text{ contains } i \text{ linearly independent points of } \Lambda\}.$$

Clearly the λ_i are nondecreasing. A very important theorem of Minkowski's (the "second theorem of Minkowski"; Theorem 11.2.2 is called the "first theorem of Minkowski") states that

$$\lambda_1 \cdots \lambda_n V(K) \le 2^n d(\Lambda).$$

W. Schmidt's great strengthening of E. Wirsing's theorem (see Section 6.1) relies very heavily on Minkowski's second theorem and on an extension of it to so-called compound convex bodies by K. Mahler.

11.2.9. Deduce Minkowski's first theorem from his second theorem.

*11.2.10. Prove Minkowski's second theorem. (Hint: Choose linearly independent vectors $\underline{w}_1, \ldots, \underline{w}_n$ so that $\underline{w}_i$ lies on the boundary of $\lambda_i K$. Pick a basis $\underline{v}_1, \ldots, \underline{v}_n$ as in Corollary 11.2.1. One can assume that $\underline{v}_i = \underline{e}_i$ for $1 \le i \le n$, without loss of generality. Thus if $\underline{x} \in \lambda_i K$ and $\underline{x} = (x_1, \ldots, x_n)$, it follows that $x_i = \ldots = x_n = 0$. Now assume that

$$\lambda_1 \cdots \lambda_n V(\tfrac{1}{2} K) > 1.$$

It will suffice to construct sets $K_1, \ldots, K_n$ such that

(i) $K_i \subseteq \lambda_i K$.

(ii) $\underline{x}, \underline{y} \in \frac{1}{2} K_i$ and $x_j = y_j$ for $1 \le j \le n$ implies that there are $\underline{x}', \underline{y}' \in \frac{1}{2} K_{i-1}$ such that $\underline{x} - \underline{y} = \underline{x}' - \underline{y}'$.

(iii) $V(\frac{1}{2} K_n) = \lambda_1 \cdots \lambda_n V(\frac{1}{2} K)$.

Then there are $\underline{x}, \underline{y} \in \frac{1}{2} K_n$ such that $\underline{x} \ne \underline{y}$ and $\underline{x} - \underline{y} \in \Lambda_0$ by Blichfeldt's theorem (Theorem 11.2.1) and condition (iii). But by condition (i) we have $K_n \subseteq \lambda_n K$, so $\underline{x}, \underline{y} \in \lambda_n(\frac{1}{2} K)$ and $\underline{x} - \underline{y} \in \lambda_n K$. Hence $x_n = y_n$. By condition (ii) there are $\underline{x}', \underline{y}' \in \frac{1}{2} K_{n-1}$ such that $\underline{x}' - \underline{y}' = \underline{x} - \underline{y}$. Also $K_{n-1} \subseteq \lambda_{n-1} K$, so $x'_{n-1} = y'_{n-1}$, $x'_n = y'_n$, etc. Ultimately we obtain $\underline{x}^{(n-1)}, \underline{y}^{(n-1)} \in \frac{1}{2} K_1$ with

$$\underline{x}^{(n-1)} - \underline{y}^{(n-1)} = \underline{x} - \underline{y}.$$

But then $x_i^{(n-1)} = y_i^{(n-1)}$ for $1 \leq i \leq n$, so $\underline{x} - \underline{y} = 0$, a contradiction.

Construct the K_i inductively. Set $K_1 = \lambda_1 K$. Say $K_1, \ldots, K_{i-1}$ have already been found. Write $R^n = \{(t_1, \ldots, t_n) | t_i \text{ is real}, 1 \leq i \leq n\}$. Let P_i be the projection of $\lambda_{i-1} K$ on the subspace $t_1 = \ldots = t_{i-1} = 0$. Choose functions $f_1 = f_1(t_i, \ldots, t_n), \ldots, f_{i-1} = f_{i-1}(t_i, \ldots, t_n)$ so that

$$(f_1, \ldots, f_{i-1}, t_i, \ldots, t_n) \in \lambda_{i-1} K$$

whenever

$$(t_i, \ldots, t_n) \in P_i.$$

Define K_i to be the set of all $\underline{u} = (u_1, \ldots, u_n)$ such that

$$\begin{aligned}
u_1 &= t_1 + (\theta_i - 1) f_1(t_i, \ldots, t_n) \\
&\vdots \\
u_{i-1} &= t_{i-1} + (\theta_i - 1) f_{i-1}(t_i, \ldots, t_n) \\
u_i &= t_i + (\theta_i - 1) t_i = \theta_i t_i \\
&\vdots \\
u_n &= t_n + (\theta_i - 1) t_n = \theta_i t_n
\end{aligned}$$

where $\theta_i = \lambda_i/\lambda_{i-1}$ and $(t_1,\ldots,t_n) \in K_{i-1}$. Now $\underline{u}/\theta_i$ is clearly in the convex set $\lambda_{i-1}K$, so $\underline{u} \in \lambda_i K$, and condition (i) follows. If $\underline{u},\underline{u}' \in K_i$ and $u_j = u_j'$ for $1 \leq j \leq n$, then $t_j = t_j'$ for $1 \leq j \leq n$, so

$$\begin{aligned}\underline{u} - \underline{u}' &= (t_1 - t_1',\ldots,t_{i-1} - t_{i-1}',0,\ldots,0)\\ &= (t_1,\ldots,t_n) - (t_1',\ldots,t_n')\end{aligned}$$

and condition (ii) follows. However, condition (iii) is rather tricky. It would seem that

$$V(K_i) = (\lambda_i/\lambda_{i-1})^{n-i+1}\, V(K_{i-1}),$$

and hence $V(K_n) = \lambda_1\ldots\lambda_n V(K)$. However, because of the rather arbitrary choice of $f_1,\ldots,f_n$, it is not even clear that the K_i are measurable!)

The preceding hint follows the paper by Davenport [22]. An ingenious method for avoiding a detailed construction of the f_i is given in the "second proof" of Bambah, Woods, and Zassenhaus [10]. Another proof is given by Cassels [18], pp. 155-160.

11.3. The Effective Dirichlet Unit Theorem

In this section we let $\underset{\sim}{K} = \underset{\sim}{Q}(\alpha)$, where $d(\underset{\sim}{K}/\underset{\sim}{Q}) = n = s + 2t$ and α is an algebraic integer with s real conjugates and 2t complex conjugates. Previously we have denoted the conjugates of α by $\alpha = \alpha_1,\alpha_2,\ldots,\alpha_n$, but here it will sometimes be more convenient to denote the real conjugates by $\alpha^{(1)},\ldots,\alpha^{(s)}$ and the pairs of complex conjugates by $\alpha^{(s+1)},\bar{\alpha}^{(s+1)};\ldots;\alpha^{(s+t)},\bar{\alpha}^{(s+t)}$. In general, if $\sigma_1,\ldots,\sigma_s;\sigma_{s+1},\bar{\sigma}_{s+1};\ldots;\sigma_{s+t},\bar{\sigma}_{s+t}$ are the corresponding isomorphisms of $\underset{\sim}{K}$ into $\underset{\sim}{C}$ that fix $\underset{\sim}{Q}$, and $\beta \in \underset{\sim}{K}$, we let $\beta^{(i)} = \sigma_i(\beta)$ and $\bar{\beta}^{(i)} = \bar{\sigma}_i(\beta)$. Also, set $r = s + t - 1$.

Definition 11.3.1. An algebraic integer η is said to be a unit if η^{-1} is also an algebraic integer.

It is easy to see that the units are precisely those algebraic integers whose minimal polynomials have the constant term +1 or -1. It is also clear (recall Definition 11.1.5) that the algebraic integer $\eta \in K$ is a unit if and only if $N(\eta) = \pm 1$. Thus the units of $\underset{\sim}{K}$ form a multiplicative Abelian group. The main result of this section, the Dirichlet unit theorem, tells us that, aside from roots of unity, this group is generated by r multiplicatively independent units.

In the following lemma and Theorems 11.3.1 and 11.3.2 we follow the presentation of Baker [7].

Lemma 11.3.1. Let $B > 0$, let $D = \prod_{1\le u<v\le n}(\alpha_u - \alpha_v)^2$, and assume that $E > |D|$ (clearly $|D| \ge 1$). Then, for $k \in \underset{\sim}{Z}^+$ and $1 \le k \le r$, there is an algebraic integer $\beta = \beta(k,B) \in K$ such that

$$|N(\beta)| \le E^{1/2}, \tag{11.3.1}$$

$$E^{1/2} \le |\beta^{(j)}| \le E \qquad (1 \le j \le r,\ j \ne k), \tag{11.3.2}$$

and

$$B/E^{1/2} \le |\beta^{(k)}| \le B. \tag{11.3.3}$$

Proof. Set

$$\beta = a_{n-1}\alpha^{n-1} + \cdots + a_1\alpha + a_0, \tag{11.3.4}$$

where $a_i \in \mathbb{Z}$ for $0 \leq i \leq n - 1$. We shall show that, given any positive real numbers $\lambda_1,\ldots,\lambda_r$, there is some $\lambda_{r+1} > 0$ such that the inequalities

$$|\beta^{(j)}| \leq \lambda_j \qquad (1 \leq j \leq s), \tag{11.3.5}$$

$$|\mathrm{Re}\ \beta^{(j)}| \leq 2^{-1/2}\lambda_j, \qquad |\mathrm{Im}\ \beta^{(j)}| \leq 2^{-1/2}\lambda_j,$$

$$(s + 1 \leq j \leq r + 1)$$

have a solution $\beta \neq 0$ of the form (11.3.4). Think of $\mathbb{R}^n$ as the cross-product of s copies of the real line with t copies of the complex plane. Then the inequalities of (11.3.5) define a compact, convex, centrally symmetric set S in $\mathbb{R}^n$ with

$$V(S) = (2\lambda_1)\cdots(2\lambda_s)(2\lambda_{s+1}^2)\cdots(2\lambda_{r+1}^2). \tag{11.3.6}$$

By (11.3.4), we have

$$\beta^{(j)} = a_{n-1}\alpha^{(j)n-1} + \cdots + a_1\alpha^{(j)} + a_0 \qquad (1 \leq j \leq s), \tag{11.3.7a}$$

$$\beta^{(j)} = a_{n-1}\alpha^{(j)n-1} + \cdots + a_1\alpha^{(j)} + a_0 \tag{11.3.7b}$$

$$(s + 1 \leq j \leq r + 1),$$

$$\bar{\beta}^{(j)} = a_{n-1}\bar{\alpha}^{(j)n-1} + \cdots + a_1\bar{\alpha}^{(j)} + a_0 \tag{11.3.7c}$$

$$(s + 1 \leq j \leq r + 1).$$

System (11.3.7) shows (roughly) that as the $(a_{n-1},\ldots,a_0)$ vector runs through all the points of the lattice Λ_0, the β vector $(\beta^{(1)},\ldots,\bar{\beta}^{(r+1)})$ runs through all the points of a lattice $\Lambda = A\Lambda_0$ with

$$d(\Lambda) = |\det A| = |D|^{1/2}$$

since the determinant is Vandermondian. Thus (roughly) if λ_{r+1} is so chosen that

$$V(S) > 2^n |D|^{1/2},$$

there will be a β vector satisfying (11.3.5) by Minkowski's theorem (Theorem 11.2.2). However, (11.3.7b) and (11.3.7c) involve complex coefficients, and Minkowski's theorem is stated only for real spaces. To escape this problem, note that the vectors

$$\underline{w}_1 = \begin{pmatrix} \gamma \\ \bar{\gamma} \end{pmatrix} \quad \text{and} \quad \underline{w}_2 = \begin{pmatrix} (\gamma - \bar{\gamma})/2i \\ (\gamma + \bar{\gamma})/2 \end{pmatrix} = \begin{pmatrix} \operatorname{Im} \gamma \\ \operatorname{Re} \gamma \end{pmatrix}$$

can be obtained from one another by row operations:

$$\underline{w}_1 \to \begin{pmatrix} \gamma \\ (\gamma + \bar{\gamma})/2 \end{pmatrix} \to \begin{pmatrix} (\gamma - \bar{\gamma})/2 \\ (\gamma + \bar{\gamma})/2 \end{pmatrix} \to \underline{w}_2.$$

By applying these operations to conjugate pairs of equations from (11.3.7b) and (11.3.7c), we eliminate all complex numbers from (11.3.7); the $\beta^{(j)}$ and $\bar{\beta}^{(j)}$ in (11.3.7b) and (11.3.7c) are replaced by $\operatorname{Im} \beta^{(j)}$ and $\operatorname{Re} \beta^{(j)}$, respectively. The matrix Λ is replaced by A' and Λ by $\Lambda' = A'\Lambda_0$, where

$$|\det A'| = 2^{-t}|\det A|.$$

Thus if

$$V(S) \geq 2^{s+t}E^{1/2}, \tag{11.3.8}$$

then

$$V(S) > 2^n 2^{-t}|D|^{1/2} = 2^n d(\Lambda'),$$

and Minkowski's theorem is applicable. We choose

$$\lambda_{r+1} = \begin{cases} E^{1/4}(\lambda_1 \cdots \lambda_s \lambda_{s+1}^2 \cdots \lambda_r^2)^{-1/2} & (t > 0) \\ E^{1/2}(\lambda_1 \cdots \lambda_{s-1})^{-1} & (t = 0) \end{cases} \tag{11.3.9}$$

to ensure (11.3.8). Hence a number $\beta \neq 0$ satisfying (11.3.5) exists. Next, by (11.3.5) and (11.3.9), we have

$$|N(\beta)| \leq E^{1/2}.$$

Since $|N(\beta)| \geq 1$, we can obtain a lower bound for $|\beta^{(j)}|$ when $1 \leq j \leq r$:

$$|\beta^{(j)}| \geq \frac{|\beta^{(j)}|}{|N(\beta)|}$$

$$\geq \begin{cases} \lambda_j(\lambda_1 \cdots \lambda_s \lambda_{s+1}^2 \cdots \lambda_{r+1}^2)^{-1} & (t > 0) \\ \lambda_j(\lambda_1 \cdots \lambda_s)^{-1} & (t = 0) \end{cases}$$

$$= \lambda_j/E^{1/2}.$$

Set

$$\lambda_j = \begin{cases} E & \text{for} \quad j \neq k \\ B & \text{for} \quad j = k \end{cases}$$

for $1 \leq j \leq r$. The result follows. □

In the next two theorems we retain the notation of Lemma 11.3.1.

Theorem 11.3.1. There are units $\eta_1,\ldots,\eta_r$ in $\underset{\sim}{K}$ such that

(11.3.10) $$|\ln|\eta_k^{(j)}|| \leq \frac{1}{2} \ln E \qquad (1 \leq j,k \leq r,\ j \neq k)$$

and

(11.3.11) $$r!\ln E \leq \ln|\eta_k^{(k)}| \leq 2(r! + 1)E^{(n+1)/2} \ln E \qquad (1 \leq k \leq r).$$

Proof. Let N_1 be the greatest integer in $E^{1/2}$. We apply Lemma 11.3.1 with $B = E^{(r!+1)\ell}$, where $\ell \in \underset{\sim}{Z}^+$ satisfies

(11.3.12) $$1 \leq \ell \leq N_1^{n+1} + 1.$$

Then $\beta = \beta(k,\ell)$, and if $\ell' > \ell$, it is clear from (11.3.3) that $\beta' = \beta(k,\ell') \neq \beta = \beta(k,\ell)$. For fixed k, there are $N_1^{n+1} + 1$ different $\beta(\ell) = \beta(k,\ell)$, and each of these has norm at most N_1. Also, each β has the form (11.3.4), and the vector $(a_{n-1},\ldots,a_0)$ can assume only N_1^n values modulo N_1. Hence by the pigeonhole principle we can find $\beta' = \beta(\ell')$ and $\beta = \beta(\ell)$ for $\ell' > \ell$ such that

(i) $\beta' \neq \beta$.

(ii) $1 \le N(\beta') = N(\beta) = N \le N_1$.

(iii) $(\beta - \beta')/N$ is an algebraic integer.

Set $\eta_k = \beta'/\beta$. Clearly $N(\eta_k) = 1$. For β an algebraic integer, $N(\beta)/\beta$ is obviously an algebraic integer, so

$$\eta_k = 1 - \left(\frac{\beta - \beta'}{N}\right)\left(\frac{N}{\beta}\right)$$

is an algebraic integer. Thus η_k is a unit. For $1 \le j \le r$ and $j \ne k$, we have

$$\begin{aligned} |\ell n|\eta_k^{(j)}|| &= \pm (\ell n|\beta^{(j)}(\ell')| - \ell n|\beta^{(j)}(\ell)|) \\ &\le \ell n\ E - \ell n\ E^{1/2} \end{aligned}$$

by (11.3.2), and (11.3.10) follows. By (11.3.12), we see that

$$\begin{aligned} \ell n|\eta_k^{(k)}| &= \ell n|\beta(\ell')| - \ell n|\beta(\ell)| \\ &\le (r! + 1)\ell'\ \ell n\ E \\ &\le 2(r! + 1)E^{(n+1)/2}\ \ell n\ E, \end{aligned}$$

and by (11.3.3) that

$$\begin{aligned} \ell n|\eta_k^{(k)}| &\ge (r! + 1)\ell'\ \ell n\ E - \frac{1}{2}\ \ell n\ E - (r! + 1)\ell\ \ell n\ E \\ &\ge r!\ \ell n\ E \end{aligned}$$

since $\ell' > \ell$. □

<u>Theorem 11.3.2</u>. Say $1 \le k,j \le r$. Then

$$\det[\ell n|\eta_k^{(j)}|] \ge \frac{3}{4}\ (r!\ \ell n\ E)^r > 0.$$

Proof. Expansion of the determinant yields a sum of r! terms. Let T_1 be the term corresponding to the main diagonal and let T_2 be any other term. At least two factors of T_2 come from outside the main diagonal, so by (11.3.10) and (11.3.11),

$$\frac{T_2}{T_1} \leq \frac{(\frac{1}{2} \ln E)^2}{(r! \ln E)^2} = \frac{1}{4(r!)^2} .$$

Hence the determinant is at least

$$T_1 - \frac{(r! - 1)T_1}{4(r!)^2} \geq \frac{3}{4} T_1 .$$

The above shows that $\underset{\sim}{K}$ has at least $r = s + t - 1$ multiplicatively independent units.

Definition 11.3.2. An algebraic number ζ is said to be a root of unity if $\zeta^m = 1$ for some $m \in \underset{\sim}{Z}^+$.

By Definition 3.1.1, roots of unity are algebraic integers; hence they are units. Results related to the following lemma have been given in Exercises 4.3.5 through 4.3.18, especially Exercise 4.3.13.

Lemma 11.3.2. If $\beta \in \underset{\sim}{K}$ is an algebraic integer and all of its conjugates lie on the unit circle, then β is a root of unity.

Proof. The powers of β have the same property, but by the proof of Theorem 11.1.2, there can be only finitely many of them. Hence $\beta^i = \beta^j$, where $i,j \in \underset{\sim}{Z}^+$ and $i > j$; that is, $\beta^{i-j} = 1$.

□

Lemma 11.3.3. If the algebraic integer α is effectively given, we can effectively list all roots of unity that lie in

$\underset{\sim}{K} = \underset{\sim}{Q}(\alpha)$. They form a finite cyclic group of even order.

Proof. All conjugates of any root of unity $\zeta \in \underset{\sim}{K}$ lie on the unit circle, so the first statement follows from Theorem 11.1.2 with $M = 2$. It is clear that the ζ form a finite group G. Since $\{1,-1\}$ is a subgroup of G, G has even order. Finally, each $\zeta \in G$ corresponds to a planar rotation by an angle $\theta = \theta(\zeta)$. Choose $\zeta' \in G$, $\zeta' \neq 1$, so that $\theta' = \theta(\zeta')$ is minimal. Then it is easy to see that every element of G is a power of ζ'.

□

We now represent each unit $\eta \in \underset{\sim}{K}$ by a vector in $\underset{\sim}{R}^r$, namely,

$$\underline{v}(\eta) = \begin{cases} (\ln|\eta^{(1)}|,\ldots,\ln|\eta^{(s)}|,\ln|\eta^{(s+1)}|^2,\ldots,\ln|\eta^{(s+t-1)}|^2), \\ \\ (\ln|\eta^{(1)}|,\ldots,\ln|\eta^{(s-1)}|), \end{cases}$$

depending on whether $t > 0$ or $t = 0$, respectively. Note that since $|N(\eta)| = 1$, the term $\ln|\eta^{(s+t)}|^2$ is linearly dependent on the components of this vector; for example, when $t > 0$,

$$\ln|\eta^{(s+t)}|^2 = -\sum_{j=1}^{s} \ln|\eta^{(j)}| - 2\sum_{j=1}^{t-1} \ln|\eta^{(s+j)}|. \tag{11.3.13}$$

Clearly $\underline{v}(\zeta) = 0$ for any root of unity ζ. It is easy to show that the set $\Lambda_U = \Lambda_U(K)$ of all such vectors $\underline{v}(\eta)$ forms a lattice. First, it is an additive group since $\underline{v}(\xi_1) + \underline{v}(\xi_2) = \underline{v}(\xi_1\xi_2)$ if ξ_1 and ξ_2 are units. Second, by the proof of Theorem 11.1.2, the set Λ_U is discrete. Third, by Theorem 11.3.2, the set Λ_U contains r linearly independent elements. Hence, by Lemma 11.2.3, it is a lattice.

Lemma 11.3.4. We can effectively determine the integer r.

Proof. It suffices to determine the integer s, and this can be done by Corollary 11.1.3 or by Sturm's theorem (see van der Waerden [123], pp. 218-222, or Exercise 11.3.1.)

□

Theorem 11.3.3. The lattice Λ_U is effective.

Proof. Let $M \in \underset{\sim}{Z}^+$ and let Γ be the set of all $\underline{v} = \underline{v}(\eta) \in \Lambda_U$ such that $\|\underline{v}\| < M$. Clearly $|\ell n|\eta^{(j)}|| < M$, $1 \leq j \leq r$, so by (11.3.13)

$$|\ell n|\eta^{(r+1)}|| < 2rM < 2nM,$$

and all the conjugates of η are bounded. Hence, by Theorem 11.1.2, we can effectively find a finite set $K_0 \subseteq \underset{\sim}{K}$ of algebraic numbers that contains all the η. We know the minimal polynomial of each element of K_0, so we can remove all the nonunits from K_0. Next, write $K_0 = K_1 \cup K_2$, where K_1 is the set of all roots of unity of K_0; this can be done by Lemma 11.3.3. By Lemma 11.3.2, every element ξ of K_2 has a conjugate whose absolute value is not unity. Define

$$S = \{\underline{0}\} \cup \{\underline{v}(\xi) \,|\, \xi \in K_2\}.$$

Then $\Gamma \subseteq S \subseteq \Lambda_U$, and S is effectively approximable and effectively bounded below by Lemma 11.1.9.

□

Theorem 11.3.4. We can effectively find units $\eta_1, \ldots, \eta_r \in \underset{\sim}{K}$ so that $\underline{v}(\eta_1), \ldots, \underline{v}(\eta_r)$ form a fundamental basis for Λ_U. Moreover, we can effectively determine upper and lower bounds on $d(\Lambda_U)$.

Proof. By using the upper bounds of (11.3.10) and (11.3.11), we can effectively find an upper bound on the determinant of Theorem 11.3.2 and hence an $N \in \mathbb{Z}^+$ such that $d(\Lambda_U) < N$. The result now follows from Lemma 11.3.4 and Corollary 11.2.4.

□

Theorem 11.3.5 (the effective Dirichlet unit theorem). One can effectively find units $\eta_1,\ldots,\eta_r \in K$ such that every unit $\eta \in K$ can be written uniquely in the form

$$\eta = \zeta\eta_1^{m_1}\cdots\eta_r^{m_r},$$

where ζ is a root of unity and $m_i \in \mathbb{Z}$ for $1 \leq i \leq r$. If η is effectively given, then ζ and the integers $m_1,\ldots,m_r$ can be effectively found.

Proof. Find $\eta_1,\ldots,\eta_r$ as in Theorem 11.3.4. If η is a unit, then $\underline{v}(\eta) \in \Lambda_U$, so

$$\underline{v}(\eta) = m_1\underline{v}(\eta_1) + \cdots + m_r\underline{v}(\eta_r),$$

where the $m_i \in \mathbb{Z}$, $1 \leq i \leq r$, are clearly unique. Thus

$$\ell n|\eta^{(j)}| = m_1\ell n|\eta_1^{(j)}| + \cdots + m_r\ell n|\eta_r^{(j)}| \qquad (1 \leq j \leq r). \tag{11.3.14}$$

Hence

$$\ell n|(\eta\eta_1^{-m_1}\cdots\eta_r^{-m_r})^{(j)}| = 0 \qquad (1 \leq j \leq r). \tag{11.3.15}$$

Since $|N(\eta)| = 1$, it immediately follows that (11.3.15) is true for $1 \leq j \leq n$, so from Lemma 11.3.2 we know that

$$\eta = \zeta\eta_1^{m_1}\cdots\eta_r^{m_r}$$

where ζ is a root of unity. Since the m_i are unique, so is ζ. For the second part of the theorem it suffices to calculate the m_i for $1 \leq i \leq r$. Since they are integers, it is enough to show that they are effectively approximable. Now (11.3.14) yields explicit formulas for the m_i by Cramer's rule, so it suffices to show that

$$|\det[\ell n|\eta_i^{(j)}|]| = d(\Lambda_U)$$

is effectively bounded below. But this follows from Theorem 11.3.4. □

Definition 11.3.3. The $\eta_1,\ldots,\eta_r$ of Theorem 11.3.5 are called a set of fundamental units.

Exercises

11.3.1. Let $P(x) \in \underset{\sim}{R}[x]$. Let $P_1(x) = P'(x)$ and define polynomials $P_2(x),\ldots,P_r(x)$ by the Euclidean algorithm as follows:

$$\begin{aligned} P(x) &= Q_1(x)P_1(x) - P_2(x) \\ P_1(x) &= Q_2(x)P_2(x) - P_3(x) \\ &\vdots \\ P_{r-2}(x) &= Q_{r-1}(x)P_{r-1}(x) - P_r(x) \\ P_{r-1}(x) &= Q_r(x)P_r(x). \end{aligned}$$

Here $d(P_{i+1}) < d(P_i)$ for $i = 1,\ldots,r - 1$. For $a \in \underset{\sim}{R}$, let $V(a;P)$ be the number of variations of sign in the sequence

$$P(a), P_1(a), \ldots, P_r(a)$$

(omit zeros; thus 5, 0, -3, 0, -1, 4 has two variations of sign). If $P(a)$ and $P(b)$ are nonzero and $a < b$, show that

$$V(a;P) - V(b;P)$$

is the number of roots (multiplicities are not counted) of P in the interval $[a,b]$. (This is known as Sturm's theorem.)

11.3.2. Find a set of fundamental units for $\underset{\sim}{Q}(\alpha)$ where (a) $\alpha^2 - 2 = 0$ and (b) $\alpha^3 + \alpha + 1 = 0$.

**11.3.3. What is the range of $d(\Lambda_U) = d(\Lambda_U(K))$ as $\underset{\sim}{K}$ varies over all algebraic number fields?

*11.3.4. Let $P(n) = \{p \mid p \text{ is a prime and } p - 1 \mid n\}$. How large is $u(n) = \Pi_{p \in P}\, p$? (Hint: The inequality

$$u(n) \leq \prod_{p \leq n+1} p \leq 4^{n+1}$$

is not best possible; one can get

$$u(n) \leq \exp(Cn^{1/2} \log^{1/2} n)$$

for some absolute constant $C > 0$ with a bit more work, but perhaps this also is much too large.)

11.3.5. (See Exercise 11.3.3.) Show that $\underset{\sim}{K} = \underset{\sim}{Q}(\alpha)$ contains at most $nu(n)$ roots of unity, where $n = d(\underset{\sim}{K}/\underset{\sim}{Q})$. (Hint: Show that the roots of unity in $\underset{\sim}{K}$ are generated by some $\zeta = \exp(2\pi i/q)$. Let $q = p_1^{e_1} \cdots p_k^{e_k}$ be the unique prime decomposition of q. Thus

$\zeta_i \in \underset{\sim}{K}$, where

$$\zeta_i = \exp(2\pi i/p_i^{e_i}).$$

Now

$$(z^{p^e} - 1)/(z^{p^{e-1}} - 1) = z^{p^{e-1}(p-1)} + z^{p^{e-1}(p-2)} + \cdots + z^{p^{e-1}} - 1$$

is irreducible by Eisenstein's criterion (consider the cases $e = 1$ and $e \geq 2$ separately), so

$$d(\zeta_i) = p_i^{e_i-1}(p_i - 1)/n.$$

Thus $p_i^{e_i-1} | n$ and $p_i - 1 | n$, so $q \leq nu(n)$.)

Chapter 12

ALGORITHMS FOR SOLVING CERTAIN DIOPHANTINE EQUATIONS

12.1. Two-Variable Homogeneous Diophantine Equations

We wish to find effectively all pairs of integers (x,y) (henceforth called solutions) such that

$$(12.1.1)\qquad f(x,y) = a_n x^n + a_{n-1} x^{n-1} y + \cdots + a_0 y^n = m,$$

where m and the a_i, $0 \leq i \leq n$, are fixed integers. This is very easy if all or all but one of the a_i is nonzero, so assume that two of the a_i are nonzero. If $m = 0$, we obtain an equation of the form

$$a_k x^k + a_{k-1} x^{k-1} y + \cdots + a_0 y^k = 0,$$

where $1 \leq k \leq n$, $a_k \neq 0$, and $a_0 \neq 0$. If $(x,y) \neq (0,0)$ is a solution then x and y are both nonzero and (px,py) is a solution for any integer p. Hence it suffices to find solutions where x and y are relatively prime. However, for such solutions, $y|a_k$ and $x|a_0$, so there are only finitely many. Thus we can assume that $m \neq 0$. Now if $a_n = 0$ in (12.1.1), then $y|m$, so there are only finitely many possibilities for y and hence only finitely many possibilities for x. Thus we can assume that $a_n \neq 0$ and (similarly) $a_0 \neq 0$.

Next we write $f(x,y) = y^n f(x/y,1)$ and factor $f(x/y,1)$ into the product of irreducible polynomials in x/y with integer coefficients; this can be done effectively by Theorem 4.1.1 and Lemma 11.1.3. Hence

$$f(x,y) = f_1(x,y)\cdots f_R(x,y), \tag{12.1.2}$$

where the $f_i(x,y)$, $1 \le i \le R$, are all homogeneous irreducible polynomials with integer coefficients. Let $(m_1,\ldots,m_R)$ be one of the finitely many R-tuples of divisors (positive or negative) of m such that $m_1\ldots m_R = m$. We can solve (12.1.1) if we can solve any system

$$f_i(x,y) = m_i \qquad (1 \le i \le R). \tag{12.1.3}$$

If $f_i(x,y) = a_1x + a_0y$ for some i, we can effectively find integers b_1,b_2,b_3,b_4 such that $x = b_1 + b_2t$ and $y = b_3 + b_4t$, where $t \in \underset{\sim}{Z}$. Then (12.1.3) reduces to a system of $R - 1$ equations in the single variable t, and the complete solution is easily found.

We next consider the special case in which all the $f_i(x,y)$ are of degree 2:

$$\begin{aligned} f_i(x,y) &= A_ix^2 + B_ixy + C_iy^2 \\ &= (4A_i)^{-1}[(2A_ix + B_iy)^2 - \Delta y^2], \end{aligned}$$

where $\Delta = \Delta_i = B_i^2 - 4A_iC_i$. Since f_i is irreducible, A_i, C_i, and Δ are all nonzero. If $\Delta < 0$, then for any solution (x,y) we have

$$|y|^2 \leq 4|\Delta|^{-1}|m_i A_i|,$$

and we are done. Thus we can assume that $\Delta_i \geq 1$ for all i, $1 \leq i \leq R$. If R = 1, we have the so-called Pell equation, which can be solved by standard techniques (see Borevich and Shafarevich [13], pp. 75-152). If $R \geq 2$, let

$$u = 2A_1 x + B_1 y$$

and

$$v = \qquad y.$$

The first two equations in (12.1.3) then have the form

$$(12.1.4) \qquad \begin{aligned} u^2 \qquad\quad - c_1 v^2 &= d_1, \\ u^2 + b_2 uv + c_2 v^2 &= d_2, \end{aligned}$$

with $d_1, d_2 \neq 0$.

Lemma 12.1.1. If the equations of (12.1.4) are not identical, $d_1 \neq 0$, and

$$h = \max(|c_1|, |d_1|, |b_2|, |c_2|, |d_2|),$$

then

$$(12.1.5) \qquad |v| \leq 36h^3.$$

Proof. By eliminating u, we find

$$(12.1.6) \qquad [b_2^2 c_1 - (c_1 + c_2)^2]v^4 + [b_2^2 d_1 + 2(d_2 - d_1)(c_1 + c_2)]v^2 - (d_2 - d_1)^2 = 0.$$

If (12.1.6) is not an identity, it follows from Lemma 3.2.1 that $|v| \leq 4 \cdot 9h^3$. If it is an identity, $d_1 = d_2$, so $b_2^2 d_1 = 0$ and $b_2 = 0$. Hence $(c_1 + c_2)^2 = 0$, and the equations are identical. □

If (12.1.5) holds, we are done. Otherwise the equations of (12.1.4) are identical, so we can replace R by R - 1; hence we have an algorithm for solving (12.1.3) in this case.

The final and most interesting case, in which one of the $f_i(x,y)$ is of degree at least 3, is settled by the following theorem:

Theorem 12.1.1 (Baker). If $f(x,y) \in \mathbb{Z}[x,y]$ is a homogeneous irreducible polynomial of degree at least 3, then for $m \in \mathbb{Z}$ the Diophantine equation

$$f(x,y) = m \tag{12.1.7}$$

has only finitely many solutions (x,y), <u>and</u> all of them can be effectively found.

The key word here is "effectively"; that the number of solutions is finite is an easy consequence of Thue's theorem of Section 5.1. In fact an effective upper bound B [in terms of m and the height and degree of f(x,y)] will be found for $\max(|x|,|y|)$.

We write (12.1.7) as

$$a_n x^n + a_{n-1} x^{n-1} y + \cdots + a_0 y^n = m. \tag{12.1.8}$$

In (12.1.8) we can assume that $a_n = 1$; multiply both sides by a_n^{n-1}, replace m by $a_n^{n-1} m$, and make the change of variable $x' = a_n x$.

Then if

$$B = B(m;n;1,a_{n-1},\dots,a_0)$$

is a bound for solutions with $a_n = 1$, the number

$$B(a_n^{n-1}m;n;1,a_n^{n-1}a_{n-1},\dots,a_n^{n-1}a_0)/|a_n|$$

is a general bound.

Now let α denote any root of $f(z,1) = 0$ and denote its conjugates by $\alpha_1 = \alpha,\alpha_2,\dots,\alpha_n$. Then

$$f(x,y) = y^n f(\frac{x}{y},1) = y^n(\frac{x}{y} - \alpha_1)\cdots(\frac{x}{y} - \alpha_n) \tag{12.1.9}$$

$$= (x - \alpha_1 y)\cdots(x - \alpha_n y) = \beta_1\cdots\beta_n,$$

where $\beta_i = x - \alpha_i y$ for $1 \le i \le n$.

Lemma 12.1.2. If for all solutions (x,y) of (12.1.7) we can effectively find two real numbers B_i,B_j depending only on m and the coefficients and degree of $f(x,y)$ such that

$$|\beta_i| < B_i \quad \text{and} \quad |\beta_j| < B_j$$

for some pair of integers i,j with $i \neq j$ and $1 \le i,j \le n$, then Theorem 12.1.1 follows.

Proof. Clearly $\beta_i - \beta_j = (\alpha_j - \alpha_i)y$, so $|y| \le (B_i + B_j)/|\alpha_j - \alpha_i|$. That $|y|$ is effectively bounded above now follows from Lemma 11.1.4, and since $|x| \le |\alpha_i||y| + B_i$, the result follows.

□

The point is that B_i, B_j must be independent of x and y. Henceforth (x,y) will always denote a pair of integers satisfying (12.1.8), so $|\beta_1 \cdots \beta_n| = |m|$. We let $\eta_1,\ldots,\eta_r$ be some set of fundamental units and $\eta_1^{(k)},\ldots,\eta_r^{(k)}$ their k-th conjugates, $1 \leq k \leq n$.

12.2. Big β and Little β

To avoid excessive proliferation of subscripts we shall denote various positive effectively computable constants in this and the following two sections by the same letter c. They may depend on m,n and the coefficients of f(x,y), but not on x or y. For example, the statement that $z \in \underset{\sim}{C}$ is effectively bounded both above and below can be written $|\ell n|z|| < c$.

Lemma 12.2.1. Integers $b_1,\ldots,b_r$ can be effectively found so that all conjugates of

$$\gamma = \beta\eta_1^{b_1}\cdots\eta_r^{b_r} \tag{12.2.1}$$

are effectively bounded both above and below.

Proof. From (12.2.1) with b_i replaced by e_i, $1 \leq i \leq r$, we obtain

$$\ell n|\gamma_k| = \ell n|\beta_k| + e_1\ \ell n|\eta_1^{(k)}| + \cdots + e_r\ \ell n|\eta_r^{(k)}|, \tag{12.2.2}$$

where (units excepted) θ_k denotes the k-th conjugate of $\theta \in \underset{\sim}{Q}(\alpha)$, $1 \leq k \leq n$. By Cramer's rule and Theorem 11.3.4, the values of the e_i that make the right-hand side of (12.2.2) zero for $1 \leq k \leq r$ are effectively approximable. Compute them to within 1/3 and replace them with integers b_i such that $|e_i - b_i| < 1$. Hence

$$|\ln|\gamma_k|| < 2(\ln|\eta_1^{(k)}| + \cdots + \ln|\eta_r^{(k)}|) \tag{12.2.3}$$
$$< c.$$

Since $r = s + t - 1$ and $2t$ of the conjugates of α are complex conjugates, it follows that $|\ln|\gamma_k||$ is effectively bounded except for at most one value of k. But since the η_i are units,

$$|\gamma_1 \cdots \gamma_n| = |N(\gamma)| = |N(\beta)| = |m|.$$

On taking logarithms, we see that there are no exceptions. □

Thus, although we still do not know if two distinct conjugates of β are bounded, we know that a number differing from β only by units (an "associate" of β) has all its conjugates bounded.

Definition 12.2.1. Let $H = \max_{i=1}^{r} |b_i|$.

Lemma 12.2.2. If $H < c$, Theorem 12.1.1 follows.

Proof. This follows from (12.2.1), Theorem 11.3.4, and Lemma 12.1.2; in fact it immediately follows from $H < c$ that all conjugates of β are bounded. □

Thus we may assume that $H > c$ finitely many times in the proof of Theorem 12.1.1.

From (12.2.1), Cramer's rule, and Theorem 11.3.4,

$$H < c \max_{i=1}^{r} |\ln| \frac{\gamma_i}{\beta_i}||. \tag{12.2.4}$$

Say the above maximum occurs at $i = k$. Then

$$(12.2.5) \qquad |\ln|\beta_k|| > c^{-1}H - c_1,$$

where $c_1 > |\ln|\gamma_k||$. Now if $H > 2cc_1$ (which we can assume; see above) the right-hand side of (12.2.5) exceeds $c^{-1}H/2$. We write simply

$$(12.2.6) \qquad |\ln|\beta_k|| > cH.$$

<u>Lemma 12.2.3</u>. There are integers k, ℓ with $1 \leq k, \ell \leq n$ depending on x and y such that

$$(12.2.7) \qquad |\beta_k| > e^{cH} \quad \text{and} \quad |\beta_\ell| < e^{-cH}.$$

<u>Proof</u>. Say $\ln|\beta_k|$ is positive (the proof is essentially the same if it is negative). Then the first inequality of (12.2.7) follows from (12.2.6). Next, since $|N(\beta)| = m$, we have

$$(n - 1) \min_{i=1}^{n} \ln|\beta_i| \leq \sum_{i \neq k} \ln|\beta_i| \leq \ln|m| - cH.$$

Say the minimum occurs at $i = \ell$. Then

$$\ln|\beta_\ell| < -cH,$$

and the result follows. □

<u>12.3. Reduction to a Problem on Linear Combinations of Logarithms</u>

The extreme behavior of the β_k and β_ℓ of Lemma 12.2.3 in terms of H will lead us to the conclusion that H is bounded. Choose an integer j different from k and ℓ, with $1 \leq j \leq n$. <u>At this point we use the assumption that</u> $n \geq 3$. Since there are only <u>two</u>

variables, we can find a relation between β_k, β_ℓ, and β_j; that is, we can find $\theta, \phi, \psi \in \underset{\sim}{Q}(\alpha)$ such that, for all x,y, we have

$$\theta(x - \alpha_k y) + \phi(x - \alpha_j y) = \psi(x - \alpha_\ell y).$$

On doing this we obtain

$$\frac{\alpha_j - \alpha_\ell}{\alpha_k - \alpha_\ell} - \frac{\beta_j}{\beta_k} = \frac{\alpha_j - \alpha_k}{\alpha_k - \alpha_\ell}\frac{\beta_\ell}{\beta_k} \neq 0. \tag{12.3.1}$$

Note that the right-hand side of (12.3.1) is very small if H is large. Multiply both sides of (12.3.1) by γ_k/γ_j, where γ is defined by (12.2.1). By Lemmas 12.2.1, 12.2.3, and 11.1.4 we find

$$0 < \left|\omega - \frac{(\eta_1^{b_1}\cdots\eta_r^{b_r})^{(k)}}{(\eta_1^{b_1}\cdots\eta_r^{b_r})^{(j)}}\right| < e^{-cH}, \tag{12.3.2}$$

where

$$\omega = \frac{\alpha_j - \alpha_\ell}{\alpha_k - \alpha_\ell}\frac{\gamma_k}{\gamma_j} \neq 0.$$

By Theorem 11.1.2, there are only finitely many possibilities for j,k,γ_j,γ_k, and these can be effectively found. Hence we can effectively write down a finite number of inequalities of the form

$$0 < |\omega^{-1}\xi_1^{b_1}\cdots\xi_r^{b_r} - 1| < e^{-cH}, \tag{12.3.3}$$

where the ξ_i are units, with the knowledge that at least one of them is true. Here ω and all of the ξ_i are effectively given.

Lemma 12.3.1. With our convention regarding c, the inequality

$$0 < |e^z - 1| < e^{-cH} \tag{12.3.4}$$

is equivalent to

$$0 < |z - 2\pi i m| < e^{-cH} \tag{12.3.5}$$

for some $m \in \mathbb{Z}$.

Proof. It is easily seen that (12.3.4) is implied by (12.3.5). To go in the other direction, assume that $0 < |e^z - 1| < \varepsilon < 1$. Choose x and y so that $-\pi < y \leq \pi$ and $z \equiv x + iy \bmod 2\pi i$. It follows immediately that $-\pi/2 \leq y \leq \pi/2$. From

$$0 < (e^x - 1)^2 < \varepsilon^2 + 2e^x(\cos y - 1)$$

we deduce that $|e^x - 1| < \varepsilon$, so

$$\ell n(1 - \varepsilon) < x < \ell n(1 + \varepsilon) \tag{12.3.6}$$

and also $1 - \varepsilon^2/2(1 - \varepsilon) < \cos y$. From $\sin^2 y + \cos^2 y = 1$ we obtain $\sin^2 y < \varepsilon^2/(1 - \varepsilon)$. But $4\pi^{-2}y^2 < \sin^2 y$, so

$$|y| < \frac{(\pi/2)\varepsilon}{(1 - \varepsilon)^{1/2}}. \tag{12.3.7}$$

If $\varepsilon < 10^{-2}$, we have $|x|, |y| < 2\varepsilon$, so for some m

$$|z - 2\pi i m| < 4\varepsilon,$$

and the result follows easily. □

From (12.3.3) and Lemma 12.3.1 it follows that

$$(12.3.8) \qquad 0 < |b_1 \log \xi_1 + \cdots + b_r \log \xi_r - 4n \log i - \log \omega| < e^{-cH},$$

where $n \in \underset{\sim}{Z}$, is a system of inequalities of which at least one is true. Note that (12.3.8) implies $|n| < cH$.

12.4. The Proof

If Feldman's theorem (Theorem 9.1.1) applied directly to the inequalities (12.3.8), we could deduce (from the true one!) that

$$(12.4.1) \qquad c_1(cH)^{-c_2} < e^{-cH},$$

and an effective upper bound on H would be immediate. (The "n" and "m" of Theorem 9.1.1 must be replaced by 1 and r + 2, respectively.) By Lemma 12.2.2, this would suffice to establish Theorem 12.1.1. Unfortunately the logarithms of (12.3.8) might be linearly dependent.

Write all the ξ_i of (12.3.8) in the canonical form provided by the effective Dirichlet unit theorem (Theorem 11.3.5). If ζ is a root of unity, $\log \zeta = (\log i)p/q$, where $p,q \in \underset{\sim}{Z}$ and $q < c$. The system of inequalities (12.3.8) now assumes the form

$$(12.4.2) \qquad 0 < |b_1' \log \eta_1 + \cdots + b_r' \log \eta_r + n' \log i + u' \log \omega| < e^{-cH},$$

where $b_i', n', u' \in \underset{\sim}{Z}$, $|u'| < c$, and $|b_i'| < cH$ for $1 \leq i \leq r$. Hence $|n'| < cH$. If for any integers $p_1, \ldots, p_{r+1}$ we have

$$(12.4.3) \qquad p_1 \log \eta_1 + \cdots + p_r \log \eta_r + p_{r+1} \log i = 0,$$

then

$$\eta_1^{p_1} \cdots \eta_r^{p_r} i^{p_{r+1}} = 1,$$

whence $p_1 = \cdots = p_r = 0$ by the Dirichlet unit theorem, and $p_{r+1} = 0$ would follow from (12.4.3). Hence $\log \eta_1, \ldots, \log \eta_r$, $\log i$ are linearly independent. Whenever ω is not a unit, its logarithm must clearly be independent of all of these, and Feldman's theorem applies to (12.4.2). When ω is a unit, we again apply Theorem 11.3.5 and obtain a system of the form (12.4.2) with $u' = 0$. Thus (12.4.1) is valid, and Theorem 12.1.1 follows.

We now have an algorithm for finding all solutions of any two-variable homogeneous Diophantine equation. This miraculous result is due to Baker [7], who in fact obtained an explicit bound for $\max(|x|, |y|)$.

Exercises

12.4.1. Show that any divisor of a homogeneous polynomial is again a homogeneous polynomial.

12.4.2. Baker [7] has shown that if $f(x,y) \in \underset{\sim}{Z}[x,y]$ is an irreducible homogeneous polynomial of degree at least 3 and $f(p,q) = m \in \underset{\sim}{Z}$, where p and q are integers, then

$$\max(|p|,|q|) < c_1 \exp(\ell n^{n+2} |m|),$$

where c_1 can be explicitly expressed in terms of the degree and the coefficients of $f(x,y)$. Use this to show that if $\alpha \in \underset{\sim}{A}$ and $n = d(\alpha) \geq 3$, then there is an effectively computable constant $c_2 = c_2(\alpha) > 0$ such that

$$(*) \qquad |\alpha - \frac{p}{q}| > c_2 q^{-n} \exp(\ell n^{1/(n+2)} q)$$

for all integers p and q, $q > 0$. (Hint: As in Baker's paper, assume first that α is a real algebraic integer and investigate the values of c_2 for which (*) is false. If $P(z)$ is the minimal polynomial of α, then

$$|q^n P(p/q)| = q^n |P(p/q) - P(\alpha)| < c_3(\alpha) q^n |\alpha - \frac{p}{q}|$$

$$< c_2 c_3 \exp(\ell n^{1/(n+2)} q).$$

Hence

$$q \leq \max(|p|,|q|) < c_1 \exp\{\ell n^{n+2} (c_2 c_3 \exp(\ell n^{1/(n+2)} q))\},$$

so, for q large,

$$(\ell n\ q - \ell n\ c_1)^{1/(n+2)} - (\ell n\ q)^{1/(n+2)} < \ell n\ 2c_2 c_3.)$$

<u>12.4.3</u>. Let $A,B \in \underset{\sim}{Z}^+$. Show that if $A^{-1}H^{-B} < e^{-H}$, then $H < A(B + 1)!$

<u>12.4.4</u>. Argue that if we know the <u>exact</u> number of solutions to a given Diophantine equation, and this number is finite, then

we have a very simple algorithm for solving it. On the other hand, explain why this algorithm no longer works if we are merely given an upper bound on the number of solutions.

*12.4.5. Let $P(x_1,\ldots,x_{h-1}) \in \mathbb{Z}[x_1,\ldots,x_{h-1}]$ be a homogeneous irreducible polynomial of degree h. Let $m \in \mathbb{Z}^+$. For what h can

$$P(x_1,\ldots,x_{h-1}) = m$$

have infinitely many integer solutions?

*12.4.6. Is there an algorithm for solving (12.1.1) if we only require that the a_i be algebraic numbers? Can we find all solutions (x,y), where x and y are algebraic integers in the field $K = \mathbb{Q}(\alpha)$, where $\alpha \in \mathbb{A}$?

Chapter 13

FURTHER RESULTS AND FURTHER PROBLEMS

In this chapter we shall indicate some of the natural directions for further study. We shall be content simply to state results without proof, but in some directions we can only offer questions.

Roth's theorem (the case $n = 1$ of Wirsing's theorem) can be used to estimate the total number of solutions of certain Diophantine equations. This was done by Davenport and Roth [24], who proved the following theorem:

<u>Theorem 13.1</u>. Let $f(x,y) = a_n x^n + a_{n-1} x^{n-1} y + \cdots + a_0 y^n$ be an irreducible polynomial of degree $n \geq 3$ with integer coefficients and let

$$g(x,y) = \sum_{i+j \leq n-3} b_{ij} x^i y^j,$$

where the b_{ij} are integers. Then the Diophantine equation

$$f(x,y) = g(x,y) \tag{13.1}$$

has less than

$$(4A)^{2n^2} B^3 + \exp(643n^2)$$

solutions, where $A = \max|a_i|$ and $B = \max|b_{ij}|$.

However, the methods of Roth seem to tell us nothing about how large the solutions are. This failing is due to the nonconstructive nature of Roth's argument; that is, the constant $c = c(\alpha,\epsilon)$ of (6.1.3) is merely shown to exist. On the other hand, Baker's methods tell us explicitly how large the solutions of many equations of the form (13.1) can be (see Baker [8], pp. 195-205).

Theorem 13.2. Say $g(x,y) \equiv m \in \underset{\sim}{Z}$, where $m \neq 0$. Then for any solution (x,y) of (13.1),

$$\max(|x|,|y|) < \exp\{(nA)^{(10n)^5} + (\log|m|)^{2n+2}\}.$$

If we are only interested in the number of solutions, Theorem 13.1 is much better. It would be a great step forward if someone could effectively compute the constant of Roth's theorem (Theorem 6.1.1), but as yet no one has done this even for Thue's theorem. In 1967 Baker ([7], p. 174, including footnote) proved the following theorem:

Theorem 13.3. Let α be an algebraic number with $n = d(\alpha) \geq 3$. Then, for p and q integers,

$$|\alpha - \frac{p}{q}| > c(\alpha)q^{-n} \exp(\ell n\ q)^{1/(n+1)},$$

where $c = c(\alpha) > 0$ can be found effectively.

These last two results were deduced by Baker not from Theorem 9.1.1 but from rather more technical results on the logarithms of algebraic numbers, tailor-made for applications to Diophantine equations.

In 1972 H. M. Stark obtained an "all-purpose" theorem on the logarithms of algebraic numbers that is sharper than any hitherto obtained:

<u>Theorem 13.4 (Stark)</u>. Let $\alpha_1,\ldots,\alpha_n;\beta_1,\ldots,\beta_n$ be algebraic numbers with $H(\alpha_i) = A_i \geq 3$ and $H(\beta_i) \leq B$. Let $\underset{\sim}{K} = \underset{\sim}{Q}(\alpha_1,\ldots,\alpha_n,\beta_1,\ldots,\beta_n)$ and $d = d(\underset{\sim}{K})$. If

$$0 < |\beta_1 \log \alpha_1 + \cdots + \beta_n \log \alpha_n| < e^{-T},$$

then either

$$T^{\log T} \leq B$$

or

$$T < C(\prod_{i=1}^{n} \log A_i)^{1+\varepsilon}$$

where $C = C(n,d,\varepsilon)$ is effectively computable.

Stark's proof uses improved estimates of units (see Siegel [112]) and discriminants.

Solving a Diophantine equation $f(x,y) = 0$ is equivalent to finding the lattice points on a certain curve, so it is natural to classify such equations according to the complexity of the corresponding curve. Experience has shown that such curves are best classified over the complex numbers, so we consider $f(x,y) = 0$ as defining y as a function g(x) of the complex variable x. Now (roughly speaking) one can make cuts on several copies of the complex plane and piece these copies together to form a smooth surface (Riemann surface) on which y is a single-valued function of x. This surface will be topologically equivalent to a sphere with $g \geq 0$ handles. This number g is called the genus of the equation $f(x,y) = 0$. For more details see Springer [118], especially pp. 1-12 and 249-299.

The methods of Baker give only fragmentary results for equations of genus $g \geq 2$, but for $g = 1$ they are essentially complete.

Definition 13.1. The polynomial $f(x_1,\ldots,x_n) \in \underset{\sim}{Z}[x_1,\ldots,x_n]$ is absolutely irreducible if it cannot be factored in any algebraic number field.

For example, if $f(x_1)$ is absolutely irreducible, it must be linear.

Theorem 13.5 (Baker and Coates [9]). Let $f(x,y)$ be an absolutely irreducible polynomial of degree n and height H. Say $f(x,y) = 0$ has genus 1. Then all integral solutions (x,y) of $f(x,y) = 0$ satisfy

$$\max(|x|, |y|) < \exp \exp \exp[(2H)^{10^{n^{10}}}].$$

The problem of finding an algorithm that solves all two-variable Diophantine equations of some fixed genus $g \geq 2$ is open. In fact, Matijasevič's [73] recent resolution of Hilbert's Tenth Problem raises the question of whether such algorithms even exist.

It is hard to resist quoting two more theorems of Baker ([2] and [7]):

Theorem 13.6. If $x,y,k \in \underset{\sim}{Z}$, where $k \neq 0$, and $y^2 = x^3 + k$, then

$$\max(|x|, |y|) < \exp[(10^{10}|k|)^{10^4}].$$

Stark later improved this to $< \exp(C|k|^{1+\epsilon})$, where $C = C(\epsilon)$ is effectively computable.

Theorem 13.7.

$$|2^{1/3} - \frac{p}{q}| > \frac{10^{-6}}{q^{2.955}}.$$

Something like Theorem 13.6 is expected because $y^2 = x^3 + k$ has genus at most 1. However, Theorem 13.7 is much stronger in form than Theorem 13.3, although it only applies to a certain fixed α. Moreover, its proof, which involves hypergeometric functions, is significantly different from that of Theorem 13.3. For more on this see the comments following Exercise 8.7.4.

Although we now know very much about the linear independence of logarithms of algebraic numbers, virtually nothing is known about the algebraic independence of such logarithms. Here the outstanding problem is Schanuel's conjecture, already discussed in Chapter 10 (take $\zeta_i = \log \alpha_i$, $1 \leq i \leq n$).

Thanks to recent work by Schmidt (see, for example, [100] or [101]) we have theorems of the Roth type for simultaneous approximation to algebraic numbers. We cite two results of this type ([100], pp. 119-120):

Theorem 13.8. Let $1, \alpha_1, \ldots, \alpha_n$ be linearly independent over $\underset{\sim}{Q}$, where $\alpha_1, \ldots, \alpha_n$ are real algebraic numbers. Let $\epsilon > 0$. Then there are only finitely many tuples of rational integers $(p_1, \ldots, p_n; q)$

$(p_1,\ldots,p_n;q)$ such that $q > 0$ and

$$|\alpha_i - \frac{p_i}{q}| < q^{-1-\frac{1}{n}-\varepsilon} \qquad (1 \le i \le n).$$

<u>Theorem 13.9</u>. Let the hypotheses of Theorem 13.8 hold. Then there are only finitely many tuples of rational integers $(q_1,\ldots,q_n;p)$ such that $q = \max(|q_1|,\ldots,|q_n|) > 0$ and

$$|\alpha_1 q_1 + \cdots + \alpha_n q_n - p| < q^{-n-\varepsilon}.$$

In fact, Schmidt has obtained a series of results both stronger and more general than these. We remark that it is an easy consequence of Exercise 9.1.3 that Theorems 13.8 and 13.9 are best possible in the sense of Exercise 6.3.4. One uses Exercise 9.1.3 with m and n replaced by 1 and n for 13.8 and by n and 1 for 13.9. These special cases of Exercise 9.1.3 are due to Dirichlet.

Perhaps the deepest question in Diophantine approximation for which there is some hope of finding a solution is the following: Is there an algebraic irrationality of degree at least 3 such that the partial quotients of its continued fraction expansion are bounded? I think it is natural to conjecture <u>no</u>, but if "bounded" is replaced by "unbounded," one still has an open question!

Finally one can ask for detailed information about the decimal expansion of π, e, $2^{1/3}$, or even $\sqrt{2}$ (see the quotation from R. Akiva at the beginning of this book).* For example, in which of these expansions does the digit 7 occur infinitely often? Here perhaps is a question beyond the range of human intelligence. But one can never know: it is said that Hilbert thought Fermat's Last Theorem would probably be solved long before the transcendence of $2^{\sqrt{2}}$ was decided. "A fence for wisdom is silence."

*Some comments about the binary expansion of $\sqrt{2}$ occur in Graham and Pollak [42]. The decimal expansion of π is carried out to 100,000 places by Shanks [105]. A startling result of Mahler's [66] asserts that $|\pi - p/q| > q^{-42}$ for integers p and q with $q \geq 2$, and E. Wirsing (unpublished) later showed that $|\pi - p/q| < q^{-21}$ has only finitely many solutions.

REFERENCES

1. J. Ax, On Schanuel's conjectures, <u>Ann. Math. 93</u> (1971), 252-268.

2. A. Baker, Rational approximations to $2^{1/3}$ and other algebraic numbers, <u>Quart. J. Math. Oxford Ser. (2) 15</u> (1964), 375-383.

3. A. Baker, Rational approximations to certain algebraic numbers, <u>Proc. London Math. Soc. 14</u> (1964), 385-398.

4. A. Baker, On some Diophantine inequalities involving the exponential function, <u>Can. J. Math. 17</u> (1965), 616-626.

5. A. Baker, Linear forms in the logarithms of algebraic numbers, I, II, and III, <u>Mathematika 13</u> (1966), 204-216; 14 (1967), 102-107; and 14 (1967), 220-228.

6. A. Baker, Simultaneous approximations to certain algebraic numbers, <u>Proc. Camb. Phil. Soc. 63</u> (1967), 693-702.

7. A. Baker, Contributions to the theory of Diophantine equations: I. On the representation of integers by binary forms, and II. The Diophantine equation $Y^2 = X^3 + k$, <u>Phil. Trans. Roy. Soc. London A263</u> (1967/68), 173-191 and 193-208.

8. A. Baker, Effective methods in Diophantine problems, Proceedings of Symposia in Pure Mathematics XX: 1969 Number Theory Institute, American Mathematical Society, Providence, R.I., 1971, pp. 195-205.

9. A. Baker and J. Coates, Integer points on curves of genus 1, Proc. Camb. Phil. Soc. 67 (1970), 595-602.

10. R. P. Bambah, A. Woods, and H. Zassenhaus, Three proofs of Minkowski's second inequality in the geometry of numbers, J. Austr. Math. Soc. 5 (1965), 453-462.

11. E. Bishop, Foundations of Constructive Analysis, McGraw-Hill, New York, 1967.

12. P. E. Blanksby and H. L. Montgomery, Algebraic integers near the unit circle, Acta Arith. 18 (1971), 355-369.

13. Z. Borevich and I. Shafarevich, Number Theory, Academic Press, New York, 1966.

14. A. Brauer, Über diophantische Gleichungen mit endlich vielen Lösungen, J. Reine Angew. Math. 160 (1929), 70-99.

15. A. D. Bryuno, The expansion of algebraic numbers in continued fractions (in Russian), Zh. Vychisl. Mat. i Mat. Fiz 4 (1964), 211-221.

16. R. C. Buck, A class of entire functions, Duke Math. J. 13 (1946), 541-559.

17. P. Bundschuh, Irrationalitätsmaße für e^a, $a \neq 0$ rational oder Liouville-Zahl, Math. Ann. 192 (1971), 229-242.

18. J. W. S. Cassels, An Introduction to Diophantine Approximation, Cambridge University Press, Cambridge, 1965.

19. J. W. S. Cassels, On a problem of Schinzel and Zassenhaus, J. Math. Sci. 1 (1966), 1-8.

20. P. J. Cohen, Set Theory and the Continuum Hypothesis, Benjamin, New York, 1966.

21. M. Cugiani, Sulla approssimabilità dei numeri algebrici mediante numeri razionali, Ann. Mat. Pura Appl. (4) 48 (1959), 135-145.

22. H. Davenport, Minkowski's inequality for the minima associated with a convex body, Quart. J. Math. Oxford 10 (1939), 119-121.

23. H. Davenport, A note on Thue's theorem, Mathematika 15 (1968), 76-87.

24. H. Davenport and K. F. Roth, Rational approximations to algebraic numbers, Mathematika 2 (1955), 160-167.

25. M. Davis, Computability and Unsolvability, McGraw-Hill, New York, 1958.

26. M. Davis, An explicit Diophantine definition of the exponential function, Comm. Pure Appl. Math. 24 (1971), 137-145.

27. M. Davis, H. Putnam, and J. Robinson, The decision problem for exponential Diophantine equations, Ann. Math. 74 (1961), 425-436.

28. L. E. Dickson, History of the Theory of Numbers, Vol. II, Chelsea, New York, 1952 (reprint).

29. R. L. Duncan, Some inequalities for polynomials, Amer. Math. Monthly 73 (1966), 58-59.

30. F. J. Dyson, The approximation to algebraic numbers by rationals, Acta Math. 79 (1947), 225-240.

31. L. Fejes Tóth, On the sum of distances determined by a pointset, Acta Math. Acad. Sci. Hungar. 7 (1956), 397-401.

32. M. Fekete and G. Szegö, On algebraic equations with integral coefficients whose roots belong to a given point set, Math. Zeit. 63 (1955), 158-172.

33. N. I. Feldman, Estimation of a linear form of logarithms of algebraic numbers (in Russian), Mat. Sb. 76 (118) (1968), 304-319.

34. N. I. Feldman, Improved estimation of a linear form of logarithms of algebraic numbers (in Russian), Mat. Sb. 77 (119) (1968), 423-436.

35. N. I. Feldman and A. B. Shidlovskii, The development and the present state of the theory of transcendental numbers, Russ. Math. Surveys 22 (1967), 1-79.

36. W. Feller, An Introduction to Probability Theory and Its Applications, Vol. I, 2nd ed., Wiley, New York, 1957.

37. A. O. Gelfond, Differenzenrechnung, Deutscher Verlag der Wissenschaften, Berlin, 1958.

38. A. O. Gelfond, The Solution of Equations in Integers, Noordhoff, Groningen, 1960.

39. A. O. Gelfond, Transcendental and Algebraic Numbers, Dover, New York, 1960.

40. A. O. Gelfond and Yu. V. Linnik, Elementary Methods in Analytic Number Theory, Rand McNally, Chicago, 1965.

41. J. Vicente Gonçalves, L'inégalité de W. Specht, Univ. Lisboa Revista Fac. Ci. A. Ci. Mat. (21) 1 (1950), 167-171.

42. R. Graham and H. Pollak, Note on a nonlinear recurrence related to $\sqrt{2}$, Math. Mag. 43 (1970), 143-145.

43. R. Güting, Polynomials with multiple zeros, Mathematika 14 (1967), 181-196.

44. G. H. Hardy and E. M. Wright, An Introduction to the Theory of Numbers, 4th ed., Clarendon Press, Oxford, 1962.

45. G. H. Hardy, J. Littlewood, and G. Pólya, Inequalities, 2nd ed., Cambridge University Press, Cambridge, 1964.

46. D. Hilbert, Mathematische Probleme, Arch. Math. Phys. 3 (1901), 44-63 and 213-237.

47. D. Hilbert, Mathematical problems, Bull. Amer. Math. Soc. 8 (1902), 437-479.

48. E. Hille, Analytic Function Theory, Vol. II, Ginn and Company, Boston, 1962.

49. F. Kasch and B. Volkman, Zur Mahlerschen Vermutung über S-Zahlen, Math. Ann. 136 (1958), 422-453.

50. J. F. Koksma, Über die Mahlersche Klasseneinteilung der transzendenten Zahlen und die Approximation komplexer Zahlen durch algebraische Zahlen, Monatsh. Math. Phys. 48 (1939), 176-189.

51. E. Landau, Note on Mr. Hardy's extension of a theorem of Mr. Pólya, Proc. Camb. Phil. Soc. 20 (1920), 14-15.

52. E. Landau, Vorlesungen über Zahlentheorie, Vol. I: Aus der elementaren und additiven Zahlentheorie, Hirzel, Leipzig, 1927.

53. E. Landau, Grundlagen der Analysis, Chelsea, New York, 1946 (reprint).

54. S. Lang, Introduction to Transcendental Numbers, Addison-Wesley, Reading, Mass., 1966.

55. S. Lang, Transcendental numbers and Diophantine approximations, Bull. Amer. Math. Soc. 77 (1971), 635-677.

56. S. Lang and H. Trotter, Continued fractions for some algebraic numbers, J. Reine Angew. Math. 225 (1972), 112-134.

57. C. G. Lekkerkerker, _Geometry of Numbers_, Wolters-Noordhoff, Groningen, 1969.

58. W. J. LeVeque, _Topics in Number Theory_, Vol. II, Addison-Wesley, Reading, Mass., 1961.

59. D. J. Lewis, ed., _Proceedings of Symposia in Pure Mathematics XX: 1969 Number Theory Institute_, American Mathematical Society, Providence, R.I., 1971.

60. L. Liouville, Sur les classes très étendus de quantités dont la valeur n'est ni algébriques, ni même réductible à des irrationelles algébriques, _Compt. Rend. Acad. Sci. (Paris) 18_ (1844), 883-885 and 910-911.

61. J. Littlewood, Mathematical notes (12): an inequality for a sum of cosines, _J. London Math. Soc. 12_ (1937), 217-221.

62. M. Loève, _Probability Theory_, 2nd ed., Van Nostrand, Princeton, N.J., 1960.

63. W. Magnus, A. Karrass, and D. Solitar, _Combinatorial Group Theory_, Interscience, New York, 1966.

64. K. Mahler, Ein Beweis des Thue-Siegelschen Satzes über die Approximation algebraischer Zahlen für binomische Gleichungen, _Math. Ann. 105_ (1931), 267-276.

65. K. Mahler, Zur Approximation der Exponentialfunktion und des Logarithmus, I and II, _J. Reine Angew. Math. 166_ (1931), 118-136; ibid., _166_ (1932), 137-150.

66. K. Mahler, On the approximation of π, Nederl. Akad. Wetensch. Proc. Ser. A56, Indag. Math. 15 (1953), 30-42.

67. K. Mahler, Lectures on Diophantine Approximations Part 1: g-adic Numbers and Roth's Theorem, University of Notre Dame, 1957.

68. K. Mahler, An application of Jensen's formula to polynomials, Mathematika 7 (1960), 98-100.

69. K. Mahler, On two extremum properties of polynomials, Illinois J. Math. 7 (1963), 681-701.

70. K. Mahler, Problem 60, Proceedings of the 1963 Number Theory Conference, University of Colorado, Boulder, 1963.

71. K. Mahler, Applications of some formulae by Hermite to the approximation of exponentials and logarithms, Math. Ann. 168 (1967), 200-227.

72. M. Marden, Geometry of Polynomials, 2nd ed., Math. Surveys No. 3, American Mathematical Society, Providence, R.I., 1966.

73. Yu. Matijasevič, Enumerable sets are Diophantine (in Russian), Dokl. Akad. Nauk SSSR 191 (1970), 279-282.

74. O. Neugebauer, Vorlesungen über Geschichte der antiken mathematischen Wissenschaften, Vol. I: Vorgriechische Mathematik, Springer, Berlin, 1934.

75. O. Neugebauer and A. Sachs, eds., Mathematical Cuneiform Texts, American Oriental Society and American Schools of Oriental Research, New Haven, Conn., 1945.

76. J. von Neumann, Ein System algebraisch unabhängiger Zahlen, Math. Ann. 99 (1928), 134-141.

77. J. von Neumann and B. Tuckerman, Continued fraction expansion of $2^{1/3}$, Math. Tables Aids Comp. 9 (1955), 23-24.

78. I. Niven, Irrational Numbers, Carus Mathematical Monograph 11, Mathematical Association of America, Wiley, New York, Rahway, 1956.

79. C. F. Osgood, The simultaneous approximation of certain k-th roots, Proc. Camb. Phil. Soc. 67 (1970), 75-86.

80. A. Ostrowski, Über ein Analogon der Wronskischen Determinante bei Funktionen mehrerer Veränderlicher, Math. Zeit. 4 (1919), 223-230.

81. A. Ostrowski, On an inequality of J. Vicente Gonçalves, Univ. Lisboa Revista Fac. Ci. A. Ci. Mat. (2) 8 (1960), 115-119.

82. J. K. Percus, Combinatorial Methods, Springer-Verlag, New York, 1971.

83. R. Péter, Recursive Functions, Academic Press, New York, 1967.

84. H. Pollard, The Theory of Algebraic Numbers, Carus Mathematical Monograph 9, Mathematical Association of America, Wiley, New York, 1950.

85. A. J. van der Poorten, On the arithmetical nature of definite integrals of rational functions, Proc. Amer. Math. Soc. 29 (1971), 451-456.

86. K. Ramachandra, A note on Baker's method, J. Austr. Math. Soc. 10 (1969), 197-203.

87. F. P. Ramsey, On a problem of formal logic. Proc. London Math. Soc. 30 (1930), 264-286.

88. R. D. Richtmyer, M. Devany, and N. Metropolis, Continued fraction expansions of algebraic numbers, Num. Math. 4 (1962), 68-84.

89. J. Robinson, Existential definability in arithmetic, Trans. Amer. Math. Soc. 72 (1952), 437-449.

90. J. Robinson, Diophantine decision problems, in Studies in Number Theory (W. J. LeVeque, ed.), Mathematical Association of America, Studies in Mathematics, Vol. 6, Prentice-Hall, Englewood Cliffs, N.J., 1969, pp. 76-116.

91. R. M. Robinson, Intervals containing infinitely many sets of conjugate algebraic integers, Studies in Mathematical Analysis and Related Topics: Essays in Honor of George Pólya, Stanford University Press, Stanford, 1962, pp. 305-315.

92. R. M. Robinson, Integer-valued entire functions, Trans. Amer. Math. Soc. 153 (1971), 451-468.

93. K. F. Roth, Rational approximations to algebraic numbers, Mathematika 2 (1955), 1-20; Corrigendum, ibid., 1968.

94. K. F. Roth, Rational approximations to algebraic numbers, Proceedings of the International Congress of Mathematicians 1958, Cambridge University Press, Cambridge, 1960, pp. 203-210.

95. W. Rudin, Real and Complex Analysis, McGraw-Hill, New York, 1966.

96. A. Schinzel, Review of a 1964 paper of Hyyrö, Zbl. Math. 137 (1967), 257-258.

97. A. Schinzel and H. Zassenhaus, A refinement of 2 theorems of Kronecker, Michigan Math. J. 12 (1965), 81-84.

98. W. Schmidt, Simultaneous approximation and algebraic independence of numbers, Bull. Amer. Math. Soc. 68 (1962), 475-478.

99. W. Schmidt, Simultaneous approximation to algebraic numbers by rationals, Acta Math. 125 (1970), 189-201.

100. W. Schmidt, Lectures on Diophantine Approximation, University of Colorado, Boulder, 1970.

101. W. Schmidt, Approximation to algebraic numbers, Enseignement Math. 17 (1971), 187-253.

102. T. Schneider, Über die Approximation algebraischer Zahlen, J. Reine Angew. Math. 175 (1936), 182-192.

103. T. Schneider, Einführung in die Transzendenten Zahlen, Springer-Verlag, Berlin, 1957.

104. I. Schur, Über die Verteilung der Wurzeln bei gewissen algebraischen Gleichungen mit ganzzahligen Koeffizienten, _Math. Zeit. 1_ (1918), 377-402.

105. D. Shanks and J. W. Wrench, Jr., Calculation of π to 100,000 decimals, _Math. Comp. 16_ (1962), 76-99.

106. A. B. Shidlovskii, On criteria for algebraic independence of values of a class of integral functions (in Russian), _Izv. Akad. Nauk SSSR Ser. Mat. 23_ (1959), 35-66.

107. C. L. Siegel, Approximation algebraischer Zahlen, _Math. Zeit. 10_ (1921), 173-213.

108. C. L. Siegel, Über einige Anwendungen diophantischer Approximationen, _Abh. Preuss. Akad. Wiss., Math.-Phys. Kl._, Nr. 1 (1929), 1-41.

109. C. L. Siegel, Die Gleichung $ax^n - by^n = c$, _Math. Ann. 114_ (1937), 57-88.

110. C. L. Siegel, _Transcendental Numbers_, Annals of Mathematics Studies, No. 16, Princeton University Press, Princeton, N.J., 1949.

111. C. L. Siegel, _Gesammelte Abhandlungen_, Band 1, Springer-Verlag, Berlin, 1966.

112. C. L. Siegel, Abschätzung von Einheiten, _Nachr. Akad. Wiss. Göttingen, Math.-Phys. Kl. 2_ (1969), 71-86.

113. W. Sierpinski, Elementary Theory of Numbers, Monograph 42, Polish Academy of Sciences, Warsaw, 1964.

114. T. Skolem, Diophantische Gleichungen, Ergebnisse der Mathematik, Vol. 5, Springer-Verlag, Berlin, 1938.

115. R. Spira, Problem E 2217, Amer. Math. Monthly 77 (1970), 192.

116. V. G. Sprindžuk, Proof of Mahler's conjecture about the measure of the set of S-numbers (in Russian), Izv. Akad. Nauk SSSR Ser. Mat. 29 (1965), 379-436.

117. V. G. Sprindžuk, Mahler's Problem in Metric Number Theory, American Mathematical Society Translations of Mathematical Monographs, No. 25, Providence, R.I., 1969.

118. G. Springer, Introduction to Riemann Surfaces, Addison-Wesley, Reading, Mass., 1957.

119. H. M. Stark, An explanation of some exotic continued fractions found by Brillhart, in Computers in Number Theory (A. O. L. Atkin and B. J. Birch, eds.), Academic Press, London, 1971, pp. 21-35.

120. P. Suppes, Axiomatic Set Theory, Van Nostrand, Princeton, N.J., 1960.

121. A. Tarski, A Decision Method for Elementary Algebra and Geometry, 2nd ed., University of California Press, Berkeley, 1951.

122. A. Thue, Über Annäherungswerte algebraischer Zahlen, *J. Reine Angew. Math. 136* (1909), 284-305.

123. B. L. van der Waerden, *Modern Algebra*, Vol. I, Ungar, New York, 1953.

124. E. Wirsing, Approximation mit algebraischen Zahlen beschränkten Grades, *J. Reine Angew. Math. 206* (1961), 67-77.

INDEX